Problems in Differential Equations

J. L. Brenner

DOVER PUBLICATIONS, INC.
Mineola, New York

Bibliographical Note

This Dover edition, first published in 2013, is an unabridged republication of the work originally published by W. H. Freeman and Company, San Francisco, in 1963.

Library of Congress Cataloging-in-Publication Data

Filippov, A. F. (Aleksei Fedorovich)
  [Sbornik zadach po differentsial?nym uravneniiam. English]
  Problems in differential equations / [translated by] J.L. Brenner.
    p. cm.
  Originally published: San Francisco : W.H. Freeman, 1963.
  Includes bibliographical references.
  ISBN-13: 978-0-486-48942-1
  ISBN-10: 0-486-48942-6
  1. Differential equations—Problems, exercises, etc. I. Brenner, J. L. (Joel Lee), 1912– II. Title.

QA371.F513 2013
515'.35076—dc23

                                                              2013014718

Manufactured in the United States by Courier Corporation
48942602   2014
www.doverpublications.com

Preface

At the beginning of every paragraph, the basic ideas needed for solving the problems that follow are given. Next, complete solutions of illustrative problems appear. For general theorems, and discussion of the more advanced material, the reader is referred at specific points to standard treatises on differential equations. Among these are

A A Andronow and C E Chaikin. Theory of Oscillations. Princeton University Press 1949

E A Coddington and N Levinson. Ordinary Differential Equations. New York: McGraw-Hill 1953 429 pp.

E Kamke. Differentialgleichungen. Lösungsmethoden und Lösungen. Band 1. Gewöhnliche Differentialgleichungen. 3 Aufl. New York: Chelsea 1948 666 pp. $9.50

N N Krasovskii. Theory of Motion. Translated by J L Brenner. Stanford University Press 1963

S Lefschetz and J LaSalle. Stability Theory of Differential Equations. Prentice-Hall 1962

V V Nemytskii and V V Stepanov. Qualitative Theory of Differential Equations. Translated under the Direction of S Lefschetz. Princeton University Press 1960

F G Tricomi. Differential Equations. Translated
   by E A McHarg. Glasgow: Blackie 1961
   50 shillings.

Other texts, of varying degrees of difficulty, are the

following:

R P Agnew. Differential Equations 2nd Ed.
   New York: McGraw-Hill 1960   485 pp. $7.50

H S Bear. Differential Equations.
   Reading, Mass: Addison-Wesley 1962   207 pp. $7.50

G Birkhoff and G C Rota. Ordinary differential equations.
   Boston: Ginn 1962   318 pp. $8.50

E A Coddington. An introduction to ordinary differential
   equations. Englewood Cliffs, New Jersey:
   Prentice-Hall 1961   292 pp. $9.00

L Collatz. Numerische Behandlung von Differentialgleichungen. Berlin: Springer 1955

S V Fagg. Differential Equations. New York: Harper
   and Row   1961   127 pp. $1.35

L R Ford. Differential Equations 2nd Ed.
   New York: McGraw-Hill 1955   291 pp. $6.50

D Greenspan. Theory and Solution of Ordinary Differential
   Equations. New York: MacMillan 1960 148 pp. $5.50

P Henrici. Discrete Variable methods in ordinary
   differential equations. New York: Wiley 1962
   407 pp. $9.95

F B Hildebrand. Introduction to Numerical Analysis.
   New York: McGraw-Hill 1956.

W Kaplan. Ordinary Differential Equations.
    Reading, Mass: Addison-Wesley 1958 534 pp. plus
    loose errata sheet. $9.75

S Lefschetz. Differential Equations: Geometric
    Theory 2nd Ed. New York: Interscience 1963
    390 pp. $10.00

C W Leininger. Differential Equations. New York:
    Harper 1962 271 pp. $6.00

W T Martin and E Reissner. Elementary Differential
    Equations. 2nd Ed Reading, Mass: Addison-Wesley
    1961 260 pp. $7.50

M Morris and O E Brown. Differential Equations 3rd Ed.
    Prentice-Hall 1952 $11.35

A L Nelson, K W Folley, M Coral. Differential Equations
    2nd Ed. Boston: Heath 1960 308 pp.

H B Phillips. Differential Equations. New York:
    Wiley 1934 125 pp.

E D Rainville. Elementary Differential Equations
    New York: Macmillan 1958 449 pp. $6.00

M Tenenbaum and H Pollard. Ordinary Differential
    Equations. New York: Harper and Row 1963
    791 pp. $10.75

L R Wilcox and H J Curtis. Elementary Differential
    Equations. Scranton, Pa: International Textbook
    Co. 1961 273 pp. $8.00

Many of the problems are routine, but each serves its spe-

cial purpose. There is a liberal sprinkling of non-routine

problems also; the more difficult ones are indicated by stars.

The explanations are so extensive that this book is almost a
complete course in itself. The only thing missing is complete
proofs of the more complicated existence and approximation
theorems.

Beginning students are sometimes handicapped by becoming
married to the use of certain letters, such as x for the
independent variable. It is good practice to translate problems
from one notation to another and back. For example, to integrate the expression $2 + x + \cos x$, change the problem to:
"integrate the expression $2 + t + \cos t$," obtain the answer
$C + \frac{1}{2} t^2 + \sin t$, and finally, report the answer
$C + \frac{1}{2} x^2 + \sin x$. This changes the marriage from one of
necessity to one of convenience.

J. L. Brenner

CONTENTS

INTRODUCTION

The term "differential equation" is a generic term that describes a certain body of subject matter. The concern of this book is almost wholly with a differential equation of the type

$$y' = f(x,y),$$

where $f(u,v)$ is a differentiable (or Lipschitzian) function of x and y where f is a differentiable function of its arguments in a certain region.

By a solution of this differential equation is meant a function $\phi(x)$ of a single independent variable that is differentiable for those values of the independent variable that lie in the given region, and such that the relation

$$\phi' = f(\phi(x),x)$$

holds for the values of x that come into question. Exact statements of this definition, and proofs of the existence of solutions, will be found in the references.

A more general type of "differential equation" is the system

$$y' = f(x,y,z), \qquad z' = g(x,y,z),$$

for which a solution is defined in a corresponding manner; in this case a solution is a pair of functions $\phi(x)$, $\psi(x)$.

Thus the set

$$y' = z$$

$$z' = - P(x)y' - Q(x)y + R(x)$$

is equivalent to the single equation

$$y'' + P(x)\ y' + Q(x)\ y = R(x)$$

in a region in which all the functions involved are defined.

In studying this page, a student will do well to rewrite the discussion in terms of a new set of letters.

THINGS TO REVIEW

Definition of derivative, and standard formulas for derivatives of elementary functions and combinations of functions, must be at hand. Some of the formulas are included at the back of this book. Certain formulas of integration are given there too; for some problems it may be convenient to have a more extensive table of integrals.

Abbreviated tables of the elementary functions are also given at the back of the book.

Section I

ISOCLINES. CONSTRUCTION OF THE DIFFERENTIAL EQUATION
FOR A FAMILY OF CURVES ISOGONAL TRAJECTORIES

If the differential equation $y' = f(x,y)$, is satisfied by a curve

going through the point (x,y), then the tangent line to the curve at the

point in question must have slope y', that is the angle α which this

line makes with the x axis must satisfy $\alpha = \arctan f(x,y)$. The geo-

metric locus of those points for which $y' = f(x,y)$ has a constant value

k is called an isocline. Thus the equations of the isoclines are

$f(x,y) = k$, where k is constant on each isocline.

To solve the differential equation $y' = f(x,y)$ geometrically, it

is sufficient to draw a number of isoclines and to sketch a curve which

crosses each isocline with the correct slope. Examples of this construc-

tion are given in any elementary text on differential equations.

The loci which intersect every curve of a given family of curves at

a constant preassigned angle ϕ are called isogonal trajectories. The

angle β which the trajectory makes with the x axis is thus ϕ units

greater or less than the angle α which the intersected trajectory makes

with the same axis: $\beta = \alpha \pm \phi$. Suppose

$$y' = f(x,y) \tag{1}$$

is the differential equation corresponding to a particular family of curves and suppose

$$y_1' = f_1(x,y) \tag{2}$$

is the differential equation of a family of isogonal trajectories. In other words, $\tan \alpha = f(x,y)$, $\tan \beta = f_1(x,y)$. Thus when equation (1) and the angle ϕ are given it is easy to write down the formula for $\tan \beta$, and thus to set up the differential equation (2) from which the isogonal trajectories are to be found.

If the equation of the given family is written in the form $F(x,y,y') = 0$, then only slight changes are needed in the above method to write down the differential equation of the isogonal trajectories. The formulas needed for the above work are

$$y' = \tan \beta = \frac{\tan \alpha \pm \tan \phi}{1 \mp \tan \alpha \tan \phi} \tag{2a}$$

Suppose the equation

$$F(x,y,C_1,\ldots,C_n) = 0 \tag{3}$$

of a family of curves is given. To construct the differential equation of this family, we differentiate equation (3) n times, and eliminate the constants $C_1,\ldots,C_n$ from the equations so obtained.

Example. Let the given family have equation

$$C_1 x + (y - C_2)^2 = 0. \tag{4}$$

5

This equation contains two parameters so we differentiate it twice as follows:

$$C_1 + 2(y - C_2)y' = 0, \qquad (5)$$

$$2y'^2 + 2(y - C_2)y' = 0. \qquad (6)$$

Let us eliminate C_1. From equation (5) we obtain $C_1 = -2(y-C_2)y'$; putting this into equation (4) we get

$$- 2xy'(y - C_2) + (y - C_2) = 0. \qquad (7)$$

Now we eliminate C_2. From equation (6) we get $y - C_2 = -\dfrac{y'^2}{y''}$; substituting this in (7), we obtain finally the differential equation we need $y' + 2xy'' = 0$.

In problems 1-14 use the method of isoclines to solve the differential equation given.

1. $y' = y - x^2.$
2. $2(y + y') = x + 3.$
3. $y' = \dfrac{x^2 + y^2}{2} - 1.$
4. $(x^2 + 1)y' = y - 2x.$
5. $yy' + x = 0.$
6. $xy' = 2y.$
7. $xy' + y = 0.$
8. $y' + 1 = 2(y - x)(y' - 1).$
9. $y'(y^2 + 1) + x = 0.$
10. $y' = \dfrac{x}{y}.$
11. $y' = \dfrac{y - 3x}{x + 3y}.$
12. $y' = \dfrac{y}{x + y}.$
13. $x^2 + y^2y' = 1.$
14. $(x^2 + y^2)y' = 4x.$

15. Give the equation of the locus of points which are maximum or minimum points of solutions of the equation $y' = f(x,y)$. What is the analytic description of a maximum or minimum point?

16. Give a method for finding the locus of the inflection points of the solutions of the equation $y' = f(x,y)$.

6

In problems 17-29 find the differential equations which correspond
to the following families of curves.

17. $y = e^{Cx}$.

18. $y = (x - C)^3$.

19. $y = Cx^3$.

20. $y = \sin(x+C)$.

21. $x^2 + Cy^2 = 2y$.

22. $y^2 + Cx = x^3$.

23. $y = C(x-C)^2$.

24. $Cy = \sin Cx$.

25. $y = ax^2 + be^x$.

26. $(x - a)^2 + by^2 = 1$.

27. $y = a \sin x + bx$.

28. $y = ax^3 + bx^2 + cx$.

29. $x = ay^2 + by + c$.

30. Write the differential equation of circles of radius 1 which
have centers on the line $y = 2x$.

31. Find the differential equation of the parabolas that have
axis parallel to the y axis and that are tangent to the two lines
$y = 0$, $y = x$.

32. Find the differential equation of the circles which are
tangent to the lines $y = 0$, $y = x$, and lie entirely in the sector
$0 \leq y \leq x$.

33. Write the differential equation of all parabolas which have
axis parallel to the y axis and pass through the origin.

34. Write the differential equation of all circles which are
tangent to the x axis.

In problems 35 and 36 find the system of differential equations which is satisfied by the given family of curves.

35. $ax + z = b$, $y^2 + z^2 = b^2$.

36. $x^2 + y^2 = z^2 - 2bz$; $y = ax + b$.

In problems 37-50 find the differential equations of the trajectories which intersect the given family isogonally at the angle ϕ :

37. $y = Cx^4$, $\varphi = 90°$.   **38.** $y^2 = x + C$, $\varphi = 90°$.
39. $x^2 = y + Cx$, $\varphi = 90°$.   **40.** $x^2 + y^2 = a^2$, $\varphi = 45°$.
41. $y = kx$, $\varphi = 60°$.   **42.** $3x^2 + y^2 = C$, $\varphi = 30°$.
43. $y^2 = 2px$, $\varphi = 60°$.   **44.** $r = a + \cos\theta$, $\varphi = 90°$.
45. $r = a\cos^2\theta$, $\varphi = 90°$.   **46.** $r = a\sin\theta$, $\varphi = 45°$.
47. $y = x\ln x + Cx$, $\varphi = \operatorname{arctg} 2$.
48. $x^2 + y^2 = 2ax$, $\varphi = 45°$.
49. $x^2 + C^2 = 2Cy$, $\varphi = 90°$. **50.** $y = Cx + C^3$, $\varphi = 90°$.

Section 2

EQUATIONS IN WHICH THE VARIABLES ARE SEPARABLE

A differential equation in which the variables are separable will take one of the forms

$$y' = f(x)\ g(y),\tag{1}$$

$$M(x)\ N(y)\ dx + P(x)\ Q(y)\ dy = 0.\tag{2}$$

To solve such an equation it is only necessary to rewrite it so that one of the variables and its differential is missing from one side and the other variable and its differential is missing from the other side.

Example. Solve the differential equation

$$x^2 y^2 y' + 1 = y.\tag{3}$$

This can be written in the form above as follows

$$x^2 y^2\ \frac{dy}{dx} = y - 1; \quad x^2 y^2 dy = (y - 1)\ dx.$$

If we divide this last equation by x^2 $(y-1)$, we obtain

$$\frac{y^2}{y - 1}\ dy = \frac{dx}{x^2}\ .$$

The variables are separated. Integrating each side separately we obtain

$$\int \frac{y^2}{y - 1}\ dy = \int \frac{dx}{x^2}\ ; \ \frac{y^2}{2} + y + \ln\left|y - 1\right| = -\frac{1}{x} + C.$$

Since we have divided $x^2(y-1)$ we must try separately each of

the solutions $x = 0$, $y - 1 = 0$. It is clear that $y = 1$ is a solu-

tion of equation (3), while $x = 0$ is not.

An equation of the form $y' = f(ax+by)$ can be transformed into

an equation with separable variables by making the substitution

$z = ax + by$, or $z = ax + by + c$, where c is an arbitrary constant.

In problems 51-65, solve the given equations and draw some

integral curves in each case. State which solutions satisfy the

initial conditions prescribed in those cases where initial conditions

are given.

51. $xydx+ (x+1)\ dy = 0$.   52. $\sqrt{y^2 + 1}\ dx = xydy$.

53. $(x^2 - 1)\ y' + 2xy^2 = 0$; $y(0) = 1$.

54. $y'\ ctg\ x +y = 2$; $y(0) = -1$.

55. $y' = 3\sqrt[3]{y^2}$; $y(2)=0$. 56. $xy'+y=y^2$; $y(1)=0.5$.
57. $2x^2yy'+y^2=2$.   58. $y'-xy^2=2xy$.
59. $e^{-s}\left(1+\dfrac{ds}{dt}\right)=1$.   60. $z'=10^{x+z}$.
61. $x\dfrac{ux}{dt}+t=1$.   62. $y'=\cos(y-x)$.
63. $y'-y=2x-3$.   64. $(x+2y)y'=1$; $y(0)=-1$.
65. $y'=\sqrt{4x+2y-1}$.

In problems 66-67 find the solution of the equations which sat-

isfies the given conditions for $x \to +\infty$.

66. $x^2 y' - \cos 2y = 1$; $y(+\infty) = \dfrac{9}{4}\pi$.

67. $3y^2y' + 16x = 2xy^3$; $y(x)$ is bounded for $x \to +\infty$

68. Find the orthogonal trajectories which correspond to the following families:

a) $y = Cx^2$; b) $y = Ce^x$; c) $Cx^2 + y^2 = 1$.

In problems 69[*] and 70[*] the variables can be separated but the resulting equations cannot be integrated by elementary functions. Thus answers to the questions posed must be obtained by a limiting process.

69[*]. Show that every interval curve of the equation

$$y' = \sqrt[3]{\frac{y^2 + 1}{x^4 + 1}}$$

has two horizontal asymptotes.

70[*]. Discuss the behavior of the integral curves of the equation

$$y' = \sqrt{\frac{\ln(1 + y)}{\sin x}}$$

in the neighborhood of the origin. Show that from every point on the boundary of the first quadrant there is an integral curve which points into this quadrant.

Section 3

GEOMETRICAL AND PHYSICAL PROBLEMS

Note: All the problems in this paragraph lead to equations
in which variables can be separated. Other exercises involving
physical problems are in Section 17.

The first step in solving the geometrical problems in this
section is to think of the required curve as having the equation
$y = y(x)$ if cartesian coordinates are to be used, and to find
the analytical form of the properties the curve is required to
have. In all problems the coordinates and slopes are denoted by
x, y, y'. The second step is to solve the differential equation
which rewrites the problem in analytical form.

In the physical problems the first step is to determine
which quantities are variables and give them names. The time,
the concentration or some other quantity is the independent variable, and the dependent variable is another quantity which varies
in a regular fashion with the independent variable. The derivative appears as the quotient of the difference $y(x + \Delta x) - y(x)$
by Δx. Sometimes it is the task of the solver to take the step

of dividing a given relation by Δx and determining the limit of this quotient as Δx approaches 0. The difference quotient represents rate of change. The derivative represents instantaneous rate. In particular, the derivative dy/dt is velocity or rate of change of the dependent variable y with respect to the time t .

In some problems the student must set up the differential equation on the basis of the physical law which is given in the text of the problem.

Example. A flask contains 10 liters of water and to it is being added a salt solution that contains 0.3 kilograms of salt per liter. This salt solution is being poured in at the rate of 2 liters per minute. The solution is being thoroughly mixed and drained off, and the mixture is drained off at the same rate so that the flask contains 10 liters at all times. How much salt is in the flask after five minutes?

Solution. The independent variable most convenient for this problem is t, and y is the dependent variable representing the number of kilograms of salt in the flask at the end of t minutes:

$$y = y(t).$$

The amount of salt added to the flask between time   t   and time

t + Δt is computed as follows.   Each minute two liters of solu-

tion is added so that in Δt   minutes, 2Δt   liters are added.

In these   2Δt   liters the amount of salt is (0.3)·   2Δt = 0.6 Δt

kilograms of salt.   On the other hand   2Δt   liters of solution

are withdrawn from the flask in an interval Δt.   Now at the

moment   t   the 10 liters in the flask contain   y(t) kilograms of

salt.   Therefore   2Δt of these liters contain   0.2 Δt ·   y(t)

kilograms of salt, if we suppose that the amount of salt does

not change in the short period of time Δt.   This is almost true

when Δt is a very short interval.   Indeed if the correct for-

mula for the last quantity is   0.2 Δt (y+α) kilograms, then

$\alpha \rightarrow$ 0 for Δt $\rightarrow$ 0.

We have computed the amount of salt added in the interval

(t,t+Δt), as well as the amount subtracted in the same interval.

But the difference between the amounts of salt present at times

t+Δt,   t   is   y(t+Δt) - y(t), and we have obtained the equation

$$y(t+\Delta t) - y(t) = 0.6\Delta t - 0.2\Delta t \cdot (y(t)+\alpha).$$

We now divide by Δt   and let Δt $\rightarrow$ 0.   The left member approaches

the derivative   y'(t), and the right member approaches 0.6 —

0.2 y(t).   The differential equation is thus:

$$y'(t) = 0.6 - 0.2\ y(t),$$

and its solution is

$$y\ (t) = 3 - Ce^{-0.2t}. \qquad (1)$$

When t is zero, the amount of salt in the flask is zero, that
is $y(0) = 0$. Equation (1) shows that when $t = 0$, we have

$$y(0) = 3 - C; \quad 0 = 3 - C; \quad C = 3.$$

The value of C is now known, so that equation (1) reads

$$y(t) = 3 - 3e^{-0.2t}.$$

To find y at the end of five minutes, we simply substitute
$t = 5$ and obtain

$$y(5) = 3 - 3e^{-0.2 \cdot 5} = 3 - 3e^{-1} \approx 1.9 \text{ kilograms of salt.}$$

71. Find a curve for which the area of the triangle determined by the tangent, the ordinate to the point of tangency and
the x axis has a constant value equal to a^2 .

72. Find a curve for which the sum of the sides of a
triangle constructed as in the previous problem has a constant
value equal to b.

73. Find a curve with the following property. The segment
of the x axis included between the tangent and normal at any
point on the curve is equal to 2a.

74. Find a curve such that the point of intersection of
an arbitrary tangent with the x axis has an abscissa half as
great as the abscissa of the point of tangency.

75. Find a curve with the following property. If through
an arbitrary point of the curve parallels are drawn to the coordinate axes and meet these axes forming a rectangle, the area
of this rectangle is divided by the curve in the ratio 1:2.

76. Find a curve such that the tangent at an arbitrary
point makes equal angles with the radius vector and the polar
axis (principal direction).

In problems 77-79 it is supposed that the amount of gas (or
liquid) contained in any fixed volume is constant. Also thorough
mixing is assumed.

77. A 20 liter vessel contains air (80% nitrogen and 20%
oxygen). 0.1 liters of nitrogen is added to the container per
second. If continual mixing takes place and material is withdrawn
at the rate at which it is added, how long will it be before the
container holds 99% nitrogen?

78. A 100 liter beaker contains 10 kilograms of salt.
Water is added at the constant rate of 5 liters per minute with
complete mixing, and drawn off at the same rate. How much salt

is in the beaker after one hour?

79. A large chamber contains 200 cubic meters of gas, 0.15% of which is carbon dioxide (CO_2). A ventilator exchanges 20 cubic meters per minute of this gas with new gas containing only 0.04% CO_2. How long will it be before the concentration of CO_2 is reduced to half its original value?

In problems 80-82 one must use Newton's law of cooling, which says that the rate of change of temperature is proportional to the difference of temperature between a cooling body and its surroundings.

80. A body cools in 10 minutes from 100° to 60°. The surroundings are at a temperature of 20°. When will the body cool to 25°?

81. A container holds 1 kilogram of water at 20°. A 0.5 kilogram mass of aluminum is added at 75°. The heat capacity of aluminum is 0.2. In 1 minute's time the water is warmed 2°. When did the aluminum cool by 1°, and when did the water warm by 1°? Assume that the water loses no heat to its surroundings. To say that the specific heat of aluminum is 0.2 means that 1 kilogram of aluminum contains as much heat as 0.2 kilograms of water.

82. A slug of metal at a temperature of 80° is put in an oven, the temperature of which is gradually warmed during an hour from a^o to b^o. Find the temperature of the metal at the end of an hour, assuming that the metal warms kT degrees per minute when it finds itself in an oven which is T degrees warmer.

83. A raft is being slowed down by resistance of the water, the resistance being proportional to the speed of the raft. If the initial speed was 1.5 meters per second and at the end of 4 seconds was 1 meter per second, when will the speed decrease to 1 centimeter per second? What total distance will the raft travel?

Problems 84-86 concern radioactive decay. The decay law states that the amount of radioactive substance that decays is proportional at each instant to the amount of substance present.

84. The strength of a radioactive substance decreases 50% in a 30-day period. How long will it take for the radioactivity to decrease to 1% of its initial value?

85. It is experimentally determined that every gram of radium loses 0.44 milligrams in 1 year. What length of time elapses before the radioactivity decreases to half its original value?

86. A certain piece of mineral contains 100 milligrams of uranium and 14 milligrams of uranium lead. It is known that uranium loses half its radioactivity in $4.5 \cdot 10^9$ years and the original amount of 238 grams of uranium decays to 206 grams of uranium lead. Calculate the mineral's age. Assume that when the mineral was born it contained no lead. Also, neglect the intermediate products of composition since the products to which uranium decomposes change themselves much more rapidly than uranium does.

87. The amount of light absorbed by a thin layer of water is proportional to the amount of incident light, and on the thickness of the layer. If a layer of water 35 cm thick absorbs half the light incident on its surface, what proportion of the incident light will be absorbed by a layer of water 200 cm thick?

In problems 88-90 it will be convenient to take the velocity to be the unknown function. Take the acceleration of gravity to be 10 m-sec^{-2}.

88. Air resistance is proportional to the square of velocity. The terminal velocity of fall of a human in air of standard density is 50 m/ sec. Neglecting the variation of air

density with altitude, find when a man's parachute should be
opened, assuming that he falls from an altitude of 1.5 km, and
his parachute must open when he reaches an altitude of 0.5 km.

89. The mass of a football is 0.4 kg. Air resists pas-
sage of the ball, the resistive force being proportional to the
square of the velocity, and being equal to 0.48 kg when the
velocity if 1 m/sec. Find the height to which the ball will
rise, and the time to reach that height, if it is thrown up-
wards with a velocity of 20 m/sec. How is the answer altered
if air resistance be neglected?

90. The football of the preceding exercise is released
(from rest) at an altitude of 16.3 m. Find its final velocity and time of fall.

In problems 91-95, assume that water emerging from an
aperture in a vessel has velocity 0.6 2gh m/sec, where
$g = 10$ m.sec^{-2} is the force of gravity, and h is the height of
the surface of the water above the aperture.

91. A vertical cylindrical vessel has diameter 2R = 1.8 m
and height H = 2.45 m. How long will it take to empty the
vessel through a hole in the bottom of diameter 2r = 6 cm?

92. Answer the same question if the axis of the cylinder is horizontal, and the hole is at a lowest point.

93. A cylindrical beaker with vertical axis can be half drained through a hole in the bottom in 5 minutes. How long would it take to empty the beaker completely?

94. A conical funnel has radius $R = 6$ cm and height $H = 10$ cm. If the opening of the funnel is a circle of diameter 0.5 cm, how long will it take to empty the entire funnel if it is initially full of water?

95. A rectangular vessel has base 60 cm by 75 cm, and height 80 cm, with an opening in the bottom 2.5 cm^2 in area. Water is being added to it at the rate of 1.8 liters per second (1 liter = 1000 cm^3.) How long does it take to fill the vessel? Compare the answer with the result that would be obtained on neglecting the hole in the bottom.

96. A uniform extensible cord 1 m long is stretched kf meters by a force of f kg. A cord of the same material meters long has mass P kg. If it is suspended by one end, how much is it extended under the weight of its own mass?

97. The density of air is 0.0012 g/cm^3. Neglecting any variation in temperature, pressure is proportional to the

density, and is 1 kg/cm^2 at the earth's surface. Find the pressure as a function of the height h.

98. A boat is held by a cable that is wound around a post, the end being held by a laborer. What is the braking force in the cable if it is wound around the post three times, the coefficient of friction between cable and post is 1/3, and the laborer exerts a force of 10 kg on the free end of the cable?

99. A closed vessel with volume v m^3 contains liquid water and air. The speed of evaporation of the water is proportional to the difference between the saturation concentration q_1 of water vapor (amount per m^3) at the given temperature, and the amount q of water vapor per m^3 actually present in the air (assume that the temperature of the air and water, and the amount of area on which evaporation occurs does not change). Initially there are m_0 grams of water in the vessel, and q_0 grams of vapor in each m^3 of air. How much water remains in the vessel at the end of t units of time?

100. The mass of a rocket, including a full chamber of fuel, is M; its net mass (without fuel) is m. The products of combustion are ejected with velocity c. If the rocket starts from rest, find Ciolkovskii's formula, which gives the

speed imparted to the rocket by the burning of the fuel, neg-

lecting the resistance of the atmosphere.

HOMOGENEOUS EQUATIONS

A homogeneous equation is one that can be written in the form $y' = f(y/x)$ or in the form $M(x,y) \; dx + N(x,y) \; dy = 0$, where $M(x,y)$ and $N(x,y)$ are homogeneous functions of the same degree. Here, a function $M(x,y)$ is called a homogeneous function of degree n, if the equation $M(kx, ky) = k^n M(x,y)$ holds for every value of k. Such an equation can be solved by making the substitution $y = tx$. This substitution reduces the equation to an equation in which the variables are separable.

Example. Solve the equation $x \; dy = (x+y) \; dx$.

This is a homogeneous equation. The substitution $y = tx$ leads to the equation $dy = tdx + xdt$. When this is substituted in the original equation, the result

$$x(xdt + tdx) = (x + tx) \; dx; \quad xdt = dx$$

is obtained. The variables are immediately separable; the equation is written

$$dt = dx/x; \quad t = \ln \left| \, x \, \right| + C.$$

Last, we introduce the original variable y, and obtain the equation $y = x(\ln |x| + C)$. Besides this solution there is a singular solution $x = 0$, which was lost when we divided by x.

An equation of the form $y' = f\left(\dfrac{a_1 x + b_1 y + c_1}{ax + by + c}\right)$ can be made homogeneous by translating the origin to the point of intersection of the lines that have equations $ax + by + c = 0$, $a_1 x + b_1 y + c_1 = 0$. If, however, these lines do not intersect the relation $a_1 x + b_1 y = k(ax + by)$ must hold. Thus the original equation has the form $y' = F(ax + by)$ and is easily converted to a form with separable variables by substituting $z = ax + by$, or $z = ax + by + c$. See Section 2.

If the number m is correctly chosen, some equations can be written in homogeneous form by making the substitution $y = z^m$. It is usually better to determine m by examining the result of the substitution. It should be remarked that this substitution is unsuccessful for the value $m = 0$.

Example. Consider the equation

$$2x^4 yy' + y^4 = 4x^6.$$

The substitution $y = z^m$ gives the new equation

$$2mx^4 z^{2m-1} z' + z^{4m} = 4x^6.$$

This equation is homogeneous if and only if all the terms have

the same degree. This would require

$$4+(2m-1) = 4m = 6.$$

The value $m = 3/2$ does satisfy both requirements. Thus the

substitution $y = z^{3/2}$ renders the original equation homogeneous.

Solve equations 101-129.

101. $(x+2y)\,dx - x\,dy = 0.$
102. $(x-y)\,dx + (x+y)\,dy = 0.$
103. $(y^2-2xy)\,dx + x^2\,dy = 0.$　　**104.** $2x^3y' = y(2x^2-y^2).$
105. $y^2 + x^2y' = xyy'.$　　**106.** $(x^2+y^2)\,y' = 2xy.$

107. $xy' - y = x\,\text{tg}\,\dfrac{y}{x}.$　　　**108.** $xy' = y - xe^{\frac{y}{x}}.$

109. $xy' - y = (x+y)\ln\dfrac{x+y}{x}.$　**110.** $xy' = y\cos\ln\dfrac{y}{x}.$

111. $(y+\sqrt{xy})\,dx = x\,dy.$　**112.** $xy' = \sqrt{x^2-y^2} + y.$
113. $(2x-4y+6)\,dx + (x+y-3)\,dy = 0.$
114. $(2x+y+1)\,dx - (4x+2y-3)\,dy = 0.$
115. $x - y - 1 + (y-x+2)\,y' = 0.$
116. $(x+4y)\,y' = 2x+3y-5.$
117. $(y+2)\,dx = (2x+y-4)\,dy.$ **118.** $y' = 2\left(\dfrac{y+2}{x+y-1}\right)^2$

119. $(y'+1)\ln\dfrac{y+x}{x+3} = \dfrac{y+x}{x+3}.$

120. $y' = \dfrac{y+2}{x+1} + \text{tg}\,\dfrac{y-2x}{x+1}.$
121. $x^3(y'-x) = y^2.$ **122.** $2x^2y' = y^3 + xy.$
123. $2x\,dy + (x^2y^4+1)\,y\,dx = 0.$
124. $y\,dx + x(2xy+1)\,dy = 0.$
125. $2y' + x = 4\sqrt{y}.$ **126.** $y' = y^2 - \dfrac{2}{x^2}.$
127. $2xy' + y = y^2\sqrt{x-x^2y^2}.$
128. $\dfrac{2}{3}xyy' = \sqrt{x^6-y^4} + y^2.$
129. $2y + (x^2y+1)\,xy' = 0.$

130. Find the trajectories which intersect the curves of

a given family at a 45° angle where the angle is measured from the tangent to the curve to the tangent to the trajectory drawn in the negative direction.

a) $y = x \ln Cx$;   b) $(x-3y)^4 = Cxy^6$.

131. Find a curve for which the tangent intersects the x axis at a point which is at the same distance from the origin as from the point of tangency.

132. Find a curve for which the distance of an arbitrary tangent from the origin is equal to the abscissa to the point of tangency.

133. For what values of α, β can or does the substitution $y = z^m$ render the differential equation $y' = ax^\alpha + bx^\beta$ homogeneous?

134.* Let the equation $f(k) = k$ have the root k_o. Show the following.

1. If $f'(k_o) < 1$, then the line $y < k x$ is tangent to no solution of the differential equation $y' = f(y/x)$.

2. If $f'(k_o) > 1$, then the line $y = k_o x$ is tangent to infinitely many solutions of the same differential equation.

135.* Find conditions which are necessary and sufficient to guarantee that every solution of the homogeneous equation $y' = f(y/x)$ is a closed curve containing the origin.

Hint. Change the equation to polar coordinates.

Section 5

LINEAR FIRST ORDER EQUATIONS

The equation

$$y' + a(x) \, y = b(x) \qquad\qquad (1)$$

is called linear. It is conveniently solved by considering

the auxiliary (reduced) equation

$$y' + a(x) \, y = 0. \qquad\qquad (2)$$

This equation is one in which the variables are separable (See

Section 2). If the constant C obtained in solving this equa-

tion is replaced by an unknown function $C(x)$, and the putative

solution is substituted in equation (1), it will be possible to

determine the unknown function so that equation (1) is satisfied.

Some equations become linear if the dependent and indepen-

dent variables are interchanged; for example, the equation

$$y = (2x + y^3) \, y'$$

is nonlinear when x is thought of as the independent variable,

but if the equation is written in the form

$$y \, dx - (2x + y^3) \, dy = 0,$$

it becomes linear because this last form is equivalent to the form

$$dx/dy - (2/y)x = y^2,$$

a linear equation with y as the independent variable. See formula (1) above.

The Bernoulli equation

$$y' + a(x)\, y = b(x)\, y^n,$$

can be converted to the standard form above by dividing by y^n and making the substitution $z = y^{1-n}$. The equation is then solved for z in terms of x. The details are given in elementary books on differential equations.

The Riccati equation, (apparently first considered by James Bernoulli, who gave fundamental algorithms for treating it)

$$y' + a(x)\, y + b(x)\, y^2 = c(x)$$

cannot always be solved by quadratures. But if one solution $y_1(x)$ is known, the substitution $y = y_1(x) + z$ converts the Ricatti equation to a Bernoulli equation and thus the solution can be completed by quadratures.

Certain equations can be written in a form free of the variable y. For example, a particular solution of the equation $y' + y^2 = x^2 - 2x$ can be found by determining the parameters a, b so that $y = ax + b$ is a solution. A second example is the following. The equation $y' + 2y^2 = 6/x^2$ has a particular solution $y = a/x$, which can be obtained by making the substitution $y = a/x$ in the equation and determining a.

Solve the equations 136-160.

136. $xy' - 2y = 2x^4.$
137. $(2x + 1)y' = 4x + 2y.$
138. $y' + y\,\text{tg}\,x = \sec x.$
139. $x(y' - y) = e^x.$
140. $x^2y' + xy + 1 = 0.$
141. $y = x(y' - x\cos x).$
142. $y' = 2x(x^2 + y).$
143. $(xy' - 1)\ln x = 2y.$
144. $xy' + (x + 1)y = 3x^2e^{-x}.$
145. $(x + y^2)\,dy = y\,dx.$
146. $(2e^y - x)y' = 1.$
147. $(\sin^2 y + x\,\text{ctg}\,y)y' = 1.$
148. $(2x + y)\,dy = y\,dx + 4\ln y\,dy.$
149. $y' = \dfrac{y}{3x - y^2}.$
150. $(1 - 2xy)y' = y(y - 1).$
151. $y' + 2y = y^2e^x.$
152. $(x + 1)(y' + y^2) = -y.$
153. $y' = y^4\cos x + y\,\text{tg}\,x.$
154. $xy^2y' = x^2 + y^3.$
155. $xy\,dy = (y^2 + x)\,dx.$
156. $xy' - 2x^2\sqrt{y} = 4y.$
157. $xy' + 2y + x^5y^3e^x = 0.$
158. $2y' - \dfrac{x}{y} = \dfrac{xy}{x^2 - 1}.$
159. $y'x^3\sin y = xy' - 2y.$
160. $(2x^2y\ln y - x)y' = y.$

Choose the appropriate independent variable in problems 161-166 so that the equation becomes linear, and solve the equation.

161. $x\,dx = (x^2 - 2y + 1)\,dy.$ **162.** $(x + 1)(yy' - 1) = y^2.$
163. $x(e^y - y') = 2.$
164. $(x^2 - 1)\,y'\sin y + 2x\cos y = 2x - 2x^3.$

165. $y(x) = \int\limits_0^x y(t)\,dt + x + 1.$

166. $\int\limits_0^x (x - t)\,y(t)\,dt = 2x + \int\limits_0^x y(t)\,dt.$

In problems 167-171, find a particular solution, rewrite the Ricatti equation as a Bernoulli equation and solve it.

167. $x^2y' + xy + x^2y^2 = 4.$ **168.** $3y' + y^2 + \dfrac{2}{x^2} = 0.$
169. $xy' - (2x + 1)\,y + y^2 = -x^2.$
170. $y' - 2xy + y^2 = 5 - x^2.$
171. $y' + 2ye^x - y^2 = e^{2x} + e^x.$

172. Find the curves which intersect the family

$$y^2 = Ce^x + x + 1.$$

orthogonally.

173. Find a curve such that the trapezoid bounded by the coordinate axes, the tangent at an arbitrary point, and the coordinate to that point has the constant area $3a^2$.

174. Find a curve such that the area of the triangle bounded by a tangent at an arbitrary point, the x axis, and the segment from the origin to the point of tangency has the constant value a^2.

175. A 100-liter beaker contains 10 kilograms of salt. Five liters of water are added to the beaker per minute and the overflow, after perfect mixing, is conducted into another 100 liter beaker which initially contained pure water. The liquid in the second beaker is also perfectly mixed. When will the amount of salt in the second beaker reach its maximum and what is the value of this maximum?

176. Let Δt be a small quantity and written as a fraction of a year. In the time Δt, every gram of radium loses 0.00044 Δt grams and yields 0.00043 Δt grams of radon. In time Δt every gram of radon loses 70 Δt grams. At the beginning of a certain experiment, there is a quantity x_o of pure radium. When will the amount of radon be a maximum?

177.* Find an approximate result indicating when the quantity of radon rises above 99% of its maximal value and when it again falls below 99% of its maximal value.

178. Find the solution of the equation

$$y' \sin 2x = 2(y + \cos x),$$

that remains bounded as $x \to \pi/2$.

179.* Show that the equation

$$\frac{dx}{dt} + x = f(t),$$

32

has a unique solution bounded for $-\infty < t < +\infty$, where $|f(t)| \leq$ M, $-\infty < t < +\infty$. Find the solution. Show further that this solution is periodic if the function $f(t)$ is periodic.

180.* Suppose $a(t) \geq c > 0$, and $f(t) \to 0$ for $t \to +\infty$. Show that every solution of the equation

$$\frac{dx}{dt} + a(t) x = f(t)$$

approaches 0 for $t \to +\infty$.

181.* In the same equation suppose that $a(t) \geq c > 0$, and let $x_o(t)$ be the solution for which the initial condition $x_o(0) = b$ is satisfied. Show that for every positive $\epsilon > 0$ there is a $\delta > 0$, such that if we perturb the function $f(t)$ and the number b by a quantity less than δ then the solution $x_o(t)$, $t \geq 0$, is perturbed by less than ϵ . The word perturbed is understood in the following sense: $f(t)$ is replaced by $f_1(t)$, and b is replaced by b_1 where

$$|f_1(t) - f(t)| < \delta , |b_1 - b| < \delta .$$

This property of the solution $x_o(t)$ is called <u>stability for persistent disturbances</u>.

182.* Let a be a positive constant, $a > 0$, and let $f(x) \to b$ for $x \to 0$. Show that exactly one solution of the equation

$$xy' + ay = f(x)$$

is bounded for $x \to 0$, and find the limit of this solution for $x \to 0$.

183.* Consider the same problem but with a a negative constant: $a < 0$. Show that every solution of the equation now has a limit, that all limits are the same as $x \to 0$. Find the value of the limit.

184.* Find a periodic solution of the equation

$$y' = y \cos^2 x + \sin x .$$

Express the solution as a definite integral.

185.* Show that only one solution of the equation

$$xy' - (2x^2 + 1) y = x^2$$

has a finite limit for $x \to +\infty$, and give the value of this limit. Express the solution as an integral.

EXACT EQUATIONS

Integrating Factors

The equation

$$M(x,y) \ dx + N(x,y) \ dy = 0 \qquad (1)$$

is called exact if the left member is the total differential

of some function $F(x,y)$. This will occur if $\partial M / \partial y = \partial N / \partial x$.

A convenient way to solve this equation is to find a function

$F(x,y)$ which has a total differential $dF(x,y) = F_x \ dx + F_y \ dy$

equal to the left member of the equation (1). Having found this

function, the general solution of equation (1) is clearly

$F(x,y) = C$, where C is an arbitrary constant.

Example. Solve the equation

$$(2x + 3x^2 y) \ dx + (x^3 - 3y^2) \ dy = 0. \qquad (2)$$

Since $\partial (2x + 3x^2 y) / \partial y = 3x^2, \ \partial (x^3 - 3y^2)/\partial x = 3x^2$,

equation (2) is exact. We must find a function $F(x,y)$ that

has a complete differential $dF = F_x \ dx + F_y \ dy$, equal to the

left member of the equation (2).

Therefore

$$F_x = 2x + 3x^2 y, \ F_y = x^3 - 3y^2. \quad (3)$$

If the first of these relations is integrated with respect to

x considering y to be a constant, we obtain

$$F = \int (2x + 3x^2 y) \ dx = x^2 + x^3 y + \phi(y),$$

where the last term is an unknown function of y. But the

partial derivative of F with respect to y is given by the

second equation (3) and from this we obtain the following

$$(x^2 + x^3 y + \phi(y))_y = x^3 - 3y^2; \ \phi'(y) = -3y^2;$$

$$\phi(y) = -y^3 + \text{const.}$$

Therefore we obtain $F(x,y) = x^2 + x^3 y - y^3$, and the general

solution of equation (2) has the form

$$x^2 + x^3 y - y^3 = C.$$

By an integrating factor of the equation

$$M(x,y) \ dx + N(x,y) \ dy = 0 \quad (4)$$

is meant a function $m(x,y) \neq 0$, so that the equation that

results when (4) is multiplied through by the function $m(x,y)$

becomes exact. Thus an integrating factor must satisfy the

relation

$$(mM)_y = (mN)_x \quad (5)$$

If M or N of equation (4) have continuous partial derivatives, and these functions are not identically 0, an integrating factor always exists. On the other hand, there is no general method for obtaining such a factor. There are recipes that are available in certain cases, as follows.

(i) If $z = \Phi(x,y)$ is a preassigned function of x and y, for example $z = x$, $z = y$, $z = xy$, $z = x/y$, it can be determined whether an integrating factor exists that depends only on z; and if one does exist, it can be found. To see this, set $m = m(z)$ in relation (5). If the result of this substitution yields an equation that can be written in the form

$$G(m, m_z, z) = 0, \qquad (6)$$

then an integrating factor depending on z alone exists and it can be found by solving equation (6).

Consider the equation

$$(y^4 - 4xy)dx + (2xy^3 - 3x^2)\, dy = 0. \qquad (7)$$

Is there an integrating factor depending only on $z = xy$?

This question is answered by substituting $m = m(z)$ in the differential condition (5):

$$[m\,(y^4 - 4xy)]_y = [m\,(2xy^3 - 3x^2)]_x$$

Since $m_x = m_z \cdot z_x = m_z \cdot y$, $m_y = m_z \cdot z_y = m_z \cdot x$, then the condition above amounts to the following

$$m_z \cdot xy \, (y^3 + x) = m \cdot 2(y^3 + x).$$

If we replace xy by z, and cancel the factor $y^3 + x$, we see that the equation to be solved is

$$z \, m_z = 2m. \tag{8}$$

Since the letters x,y do not appear explicitly in this equation, an integrating factor exists which depends only on z. From equation (8) we find the result $m = Cz^2$. But the constant C is arbitrary and can be taken to be 1. Thus $m = x^2 y^2$ is an integrating factor for equation (7).

If equation (4) can be written in the form

$$d \, \phi(x,y) + M_1(x,y) \, dx + N_1(x,y) \, dy = 0, \tag{9}$$

where $d \, \phi(x,y)$ is the total differential of some function $\phi(x,y)$ then the equation can be solved if we can find, according to the recipe above, an integrating factor for the subsidiary equation $M_1(x,y) \, dx + N_1(x,y) \, dy = 0$, that is a function of z alone $[z = \phi(x,y)]$. If such an integrating factor exists, then it is also an integrating factor for equation (9).

Example. Consider the equation

$$(xy + y^4) \, dx + (x^2 - xy^3) \, dy = 0.$$

If we collect the terms according to degree, we obtain

$$x(\, y \, dx + x \, dy) + y^3 \, (y \, dx - x \, dy) = 0.$$

Now we divide by x and obtain the equation

$$d \, (xy) + \frac{y^3}{x} \, (y \, dx - x \, dy) = 0, \qquad (10)$$

in which the first term is a total differential. The subsidiary equation is

$$\frac{y^3}{x} \, (y \, dx - x \, dy) = 0 \qquad (11)$$

and has an integrating factor depending only on $z = xy$, as the above method shows. This factor is $m = 1/z^2$.

Now we multiply the entire equation (10) by this integrating factor and obtain the equation

$$\frac{d(xy)}{x^2 y^2} + \frac{y^2}{x^3} \, dx - \frac{y}{x^2} \, dy = 0,$$

in which the left member is a total differential.

Sometimes parts of equation (4) can be recognized as total differentials of a function $\phi(x,y)$. In such a case the equation may be simplified by introducing new variables (x,z) or (y,z) in place of the variables (x,y). Here z equals $\phi(x,y)$.

In special cases it is even desirable to change the variables from (x,y) to (u,v), where $u = \phi (x,y)$ and $v = \psi(x,y)$.

Example. We consider equation (10) from this point of view. Since $y\, dx - x\, dy = y^2\, d\ (x/y)$, the equation (10) can be written in the form

$$d\ (xy) + \frac{y^5}{x}\, d\ (x/y) = 0.$$

We now write $xy = u$, $x/y = v$, and the equation takes the form $du + u^2 dv/v^3 = 0$, which is easily solved since the variables are separable.

Note that if an exact equation has solution $F = C$, then any function of F is an integrating factor. If

$$Mdx + Ndy$$

has integrating factor $\phi(z)$ and solution $F = C$, then $\psi(F)\ \phi(z)$ is an integrating factor, where ψ is an arbitrary differentiable function.

In problems 186-194, check that the differential equations

are exact and solve each equation.

186. $2xy\,dx + (x^2 - y^2)\,dy = 0.$

187. $(2 - 9xy^2)\,x\,dx + (4y^2 - 6x^3)\,y\,dy = 0.$

188. $e^{-y}\,dx - (2y + xe^{-y})\,dy = 0.$

189. $\frac{y}{x}\,dx + (y^3 + \ln x)\,dy = 0.$

190. $\frac{3x^2 + y^2}{y^2}\,dx - \frac{2x^3 + 5y}{y^3}\,dy = 0.$

191. $2x\left(1 + \sqrt{x^2 - y}\right)dx - \sqrt{x^2 - y}\,dy = 0.$

192. $(1 + y^2 \sin 2x)\,dx - 2y \cos^2 x\,dy = 0.$

193. $3x^2(1 + \ln y)\,dx = \left(2y - \frac{x^3}{y}\right)dy.$

194. $\left(\frac{x}{\sin y} + 2\right)dx + \frac{(x^2 + 1)\cos y}{\cos 2y - 1}\,dy = 0.$

Solve equations 195-220 by finding an integrating factor

of suitable form, or by making a suitable change of variables.

195. $(x^2 + y^2 + x)\,dx + y\,dy = 0.$

196. $(x^2 + y^2 + y)\,dx - x\,dy = 0.$

197. $x\,dx = (x\,dy + y\,dx)\sqrt{1 + x^2}.$

198. $xy^2(xy' + y) = 1.$

199. $y^2\,dx - (xy + x^3)\,dy = 0.$

200. $\left(y - \frac{1}{x}\right)dx + \frac{dy}{y} = 0.$

201. $(x^2 + 3 \ln y)\,y\,dx = x\,dy.$

202. $y^2\,dx + (xy + \operatorname{tg} xy)\,dy = 0.$

203. $y(x + y)\,dx + (xy + 1)\,dy = 0.$

204. $y(y^2 + 1)\,dx + x(y^2 - x + 1)\,dy = 0.$

205. $(x^2 + 2x + y)\,dx = (x - 3x^2y)\,dy.$

206. $y\,dx - x\,dy = 2x^3 \operatorname{tg} \frac{y}{x}\,dx.$

207. $y^2\,dx + (e^x - y)\,dy = 0.$

208. $xy\,dx = (y^3 + x^2y + x^2)\,dy.$

209. $x^2y(y\,dx + x\,dy) = 2y\,dx + x\,dy.$

210. $(x^2 - y^2 + y)\,dx + x(2y - 1)\,dy = 0.$

211. $(2x^2y^2 + y)\,dx + (x^3y - x)\,dy = 0.$

212. $(2x^2y^3 - 1)\,y\,dx + (4x^2y^3 - 1)\,x\,dy = 0.$

213. $y(x + y^2)\,dx + x^2(y - 1)\,dy = 0.$

214. $(x^2 - \sin^2 y)\,dx + x \sin 2y\,dy = 0.$

215. $x(\ln y + 2 \ln x - 1)\,dy = 2y\,dx.$

216. $(x^2 + 1)(2x\,dx + \cos y\,dy) = 2x \sin y\,dx.$

217. $(2x^3y^2 - y)\,dx + (2x^2y^3 - x)\,dy = 0.$

218. $x^2y^3 + y + (x^3y^2 - x)\,y' = 0.$

219. $(x^2 - y)\,dx + x(y + 1)\,dy = 0.$

220. $y^2(y\,dx - 2x\,dy) = x^3(x\,dy - 2y\,dx).$

Section 7

QUESTIONS ON EXISTENCE AND UNIQUENESS OF SOLUTIONS

A differential equation, particularly a differential
equation of first order, is easily rewritten as an integral
equation. In this manner a solution with given initial conditions can be found by the method of successive approximations.

Estimate the length of the interval on which Picard's
theorem guarantees the existence of a solution and the convergence of his method of successive approximations [Coddington-
Levinson, p. 6, (1.6)]

221. $y' = x - y^2$, $y(0)$. Find y_0, y_1, y_2, y_3.

222. $y' = y^2 - 3x^2 - 1$, $y(0) = 1$. Find y_0, y_1, y_2.

223. $y' = y + e^y$, $y(0) = 1$. Find y_0, y_1, y_2.

224.* Estimate the error of the approximation y_3 in
problem 221 for $x = 0.5$ and for $x = 1$.

Hint. Calculate the remainder of the series that is used
to prove the existence of a solution.

225.* Show that the solution of the equation $y' = x^3 - y^3$
with arbitrary initial value $y(x_0) = y_0$ exists for all x,
$x_0 \leq x < \infty$.

In problems 226-240 state a theorem which guarantees the uniqueness of the solution of the equation in question. Be sure that the example satisfies the hypotheses of the theorem. In problems 229 and 230 the right member of the equation is defined at $y = 0$ by the specification that it be continuous.

226. $y' = 2xy + y^2$.

227. $y' = 3\sqrt[9]{y^2}$.

228. $y' = 3\sqrt[3]{y^2} + 1$.

229. $y' = y \ln y$.

230. $y' = y \ln^2 y$.

231. $y' = \sqrt[3]{y} + x$.

232. $y' = \dfrac{y+2}{x+y}$.

233. $y' = \dfrac{x+2y-4}{x-y-1}$.

234. $y' = \operatorname{tg} y + 1$.

235. $y' = \sqrt{\sin y}$.

236. $y' = 2 + \sqrt[3]{y-2x}$.

237. $y' = \sqrt{x+2y} - x$.

238. $y' = \sin x + \cos y$.

239. $y' = \dfrac{\sqrt{y}-x}{x-2}$.

240. $xy' = y + \sqrt{y^2 - x^2}$.

Section 8

EQUATIONS IN WHICH THE DERIVATIVE APPEARS IMPLICITLY

An equation of the form $F(x,y,y') = 0$ can be solved in the following manner.

If it is possible to solve the equation $F(x,y,y') = 0$ explicitly for y' in terms of x and y, the problem is re-written in the form $y' = f(x,y)$. This equation must then be solved. In some cases more than one equation $y' = f(x,y)$ is obtained from the given equation. In this case each equation must be solved and the corresponding solutions checked in the original equation.

A method called parametric solution proceeds as follows. (We explain a simple variant of a method described more gen-erally in standard texts.)

The first step is to solve the equation $\bar{F}(x,y,y') = 0$ for y, obtaining a result of the form $y = f(x,y')$. If we introduce the parameter

$$p = \frac{dy}{dx} = y',\tag{1}$$

we obtain

$$y = f(x,p). \tag{2}$$

Now we take the total derivative of equation (2), replace
dy by p dx (see (1)), and obtain an equation of the form

$$M(x,p) \; dx + N \; (x,p) \; dp = 0.$$

If this equation can be solved in the form $x = \phi(p)$, the original equation is solved in parametric form by use of equation
(2): $x = \phi(p)$, $y = f(\phi(p), \; p)$.

The equation $x = f(y,y')$ can be solved in the same manner.

This method is to be used in problems 267-286 below.

A solution $y = \phi(x)$ of the equation $F(x,y,y') = 0$ is
called singular, if every point of this curve has the property
that a second solution of the differential equation passes
through this point, that is, has the same slope as the solution
$y = \phi(x)$, and yet the second solution is different from the
singular solution in every neighborhood of the given point.
(The term singular solution is used in other senses by certain
authors.)

If the function $F(x,y,y')$ and its partial derivatives $\partial F/\partial y$, $\partial F/ y'$ are continuous, then every singular solution of the equation

$$F(x,y,y') = 0 \qquad (3)$$

also satisfies the condition

$$\frac{\partial F(x,y,y')}{y'} = 0. \qquad (4)$$

Thus the singular solutions of equation (3) are obtained by eliminating y' from equations (3), (4). The equation $\psi(x,y) = 0$ obtained in this elimination is called the discriminant of the original differential equation. It is necessary to check each branch of the discriminant curve to find if it is a valid solution of equation (3). If such is the case then it can be checked further whether it is a singular solution, that is, whether more than one integral curve passes through each of its points. This last property is easily checked whenever the complete or general solution of the equation is known.

If a family of curves $\vartheta(x,y,C) = 0$ has an envelope $y = \phi(x)$, and if the family satisfies the equation $F(x,y,y') = 0$, then the envelope will be a singular solution of the equation. If the function ϑ has continuous derivatives with respect to C, a necessary condition for the existence of an envelope is that the relations

$$\vartheta(x,y,C) = 0, \qquad \frac{\partial \vartheta(x,y,C)}{\partial C} = 0$$

hold. Thus, one can check whether the curve obtained by eliminating these two curves is indeed an envelope of the given family.

In problems 241-250 find all solutions of the given equations; indicate the singular solutions if there are any; draw pictures.

241. $y'^3 - y^2 = 0$.

242. $8y'^3 = 27y$.

243. $(y' + 1)^3 = 27(x+y)^2$.

244. $y^2(y'^2 + 1) = 1$.

245. $y'^2 - 4y^3 = 0$.

246. $y'^2 = 4y^3(1 - y)$.

247. $xy'^2 = y$.

248. $yy'^3 + x = 1$.

249. $y'^3 + y^2 = yy'(y'+1)$.

250. $4(1-y) = (3y-2)^2 y'^2$.

Solve equations 251-266 for y' and then find the solution by the usual methods. Indicate the singular solutions if there are any.

251. $y'^2 + xy = y^2 + xy'$.

252. $xy'(xy' + y) = 2y^2$.

253. $xy'^2 - 2yy' + x = 0$.

254. $xy'^2 = y(2y' - 1)$.

255. $y'^2 + x = 2y$.

256. $y'^3 + (x + 2)e^y = 0$.

257. $y'^2 - 2xy' = 8x^2$.

258. $(xy' + 3y)^2 = 7x$.

259. $y'^2 - 2yy' = y^2(e^x - 1)$.

260. $y'(2y - y') = y^2 \sin^2 x$.

261. $y'^4 + y^2 = y^4$.

262. $x(y - xy')^2 = xy'^2 - 2yy'$.

263. $y(xy' - y)^2 = y - 2xy'$.

264. $yy'(yy' - 2x) = x^2 - 2y^2$.

265. $y'^2 + 4xy' - y^2 - 2x^2y = x^4 - 4x^2$.

266. $y(y - 2xy')^2 = 2y'$.

Solve equations 267-286 in parametric form.

267. $x = y'^3 + y'$.

268. $x(y'^2 - 1) = 2y'$.

269. $x = y'\sqrt{y'^2 + 1}$.

270. $y'(x - \ln y') = 1$.

271. $y = y'^2 + 2y'^3$.

272. $y = \ln(1 + y'^2)$.

273. $(y' + 1)^3 = (y' - y)^2$.

274. $y = (y' - 1)e^{y'}$.

275. $y'^4 - y'^2 = y^2$.

276. $y'^2 - y'^3 = y^2$.

277. $y'^4 = 2yy' + y^2$.

278. $y'^2 - 2xy' = x^2 - 4y$.

279. $5y + y'^2 = x(x + y')$.

280. $x^2 y'^3 = xyy' + 1$.

281. $y'^3 + y^2 = xyy'$.

282. $2xy' - y = y' \ln yy'$.

283. $y' = e^{\frac{xy'}{y}}$.

284. $y = xy' - x^2 y'^3$.

285. $v = 2xy' + y^2 y'^3$.

286. $y(y - 2xy')^3 = y'^2$.

Equations 287-297 are of the Lagrange-Clairaut type.

287. $y = xy' - y'^2$.

288. $y + xy' = 4\sqrt{y'}$.

289. $y = 3xy' - 7y'^3$.

290. $y = xy' - (2 + y')$.

291. $x(y'^2 + 1) = 2yy'$.

292. $y = xy'^2 - 2y'^3$.

293. $xy' - y = \ln y'$.

294. $xy'(y' + 2) = y$.

295. $2y'^2(y - xy') = 1$.

296. $2xy' - y = \ln y'$.

297. $y'^3 = 3(xy' - y)$.

298. Find a curve such that its tangent cuts off a triangle of area $2a^2$ on the coordinate axes.

299. Find a curve such that the sum of the squares of the reciprocals of the lengths of the segments cut by its tangent from the coordinate axes is 1.

300. Find a curve that goes through the origin, and is so arranged that the segments of the normal cut off in the first quadrant, has a constant length equal to 2.

Section 9

MISCELLANEOUS FIRST ORDER EQUATIONS

Solve equations 301-330 and draw a graph of the solutions.

All problems in this section use methods introduced earlier.

301. $xy' + x^2 + xy - y = 0.$ **302.** $2xy' + y^2 = 1.$

303. $(2xy^2 - y)\,dx + x\,dy = 0.$

304. $(xy' + y)^2 = x^2 y'.$ **305.** $y - y' = y^2 + xy'.$

306. $(x + 2y^3)\,y' = y.$ **307.** $y'^3 - y' e^{2x} = 0.$

308. $x^2 y' = y(x + y).$ **309.** $(1 - x^2)\,dy + xy\,dx = 0.$

310. $y'^2 + 2(x - 1)\,y' - 2y = 0.$

311. $y + y' \ln^2 y = (x + 2\ln y)\,y'.$

312. $x^2 y' - 2xy = 3y.$ **313.** $x + yy' = y^2 \left(1 + y'^2\right).$

314. $y = (xy' + 2y)^2.$ **315.** $y' = \dfrac{1}{x - y^2}.$

316. $y'^3 + (3x - 6)\,y' = 3y.$ **317.** $x - \dfrac{y}{y'} = \dfrac{2}{y}.$

318. $2y'^3 - 3y'^2 + x = y.$ **319.** $(x + y)^2\,y' = 1.$

320. $2x^3 yy' + 3x^2 y^2 + 7 = 0.$ **321.** $\dfrac{dx}{x} = \left(\dfrac{1}{y} - 2x\right)dy.$

322. $xy' = e^y + 2y'.$ **323.** $2(x - y^2)\,dy = y\,dx.$

324. $x^2 y'^2 + y^2 = 2x(2 - yy').$

325. $dy + (xy - xy^3)\,dx = 0.$ **326.** $2x^2 y' = y^2(2xy' - y).$

327. $\dfrac{y - xy'}{x + yy'} = 2.$ **328.** $x(x - 1)\,y' + 2xy = 1.$

329. $xy(xy' - y)^2 + 2y' = 0.$ **330.** $(1 - x^2)\,y' - 2xy^2 = xy.$

Solve equations 331-420.

331. $y' + y = xy^3$.

332. $(xy^2 - x) dx + (y + xy) dy = 0$.

333. $(\sin x + y) dy + (y \cos x - x^2) dx = 0$.

334. $3y'^3 - xy' + 1 = 0$.   **335.** $yy' + y^2 \operatorname{ctg} x = \cos x$.

336. $(e^y + 2xy) dx + (e^y + x) x \, dy = 0$.

337. $xy'^2 = y - y'$.   **338.** $x(x+1)(y'-1) = y$.

339. $y(y - xy') = \sqrt{x^4 + y^4}$.   **340.** $xy' + y = \ln y'$.

341. $x^2 (dy - dx) = (x + y) y \, dx$.

342. $y' + x\sqrt[v]{v} = 3v$.   **343.** $(x \cos y + \sin 2v) v' = 1$.

344. $y'^2 - yy' + e^x = 0$.   **345.** $y' = \dfrac{x}{y} e^{2x} + y$.

346. $(xy' - y)^3 = y'^3 - 1$.   **347.** $(4xy - 3) y' + y^2 = 1$.

348. $y'\sqrt{x} = \sqrt{y - x} + \sqrt{x}$.   **349.** $xy' = 2\sqrt{y} \cos x - 2y$.

350. $3y'^4 = y' + y$.   **351.** $y^2 (y - xy') = x^3 y'$.

352. $y' = (4x + y - 3)^2$.

353. $(\cos x - x \sin x) y \, dx + (x \cos x - 2y) dy = 0$.

354. $x^2 y'^2 - 2xyy' = x^2 + 3y^2$.

355. $\dfrac{xy'}{y} + 2xy \ln x + 1 = 0$.   **356.** $xy' = x\sqrt{y - x^2} + 2y$.

357. $(1 - x^2 y) dx + x^2 (y - x) dy = 0$.

358. $(2xe^y + y^4) y' = ye^y$.   **359.** $xy' (\ln y - \ln x) = y$.

360. $2y' = x + \ln y'$.

361. $(2x^2 y - 3y^2) y' = 6x^2 - 2xy^2 + 1$.

362. $yy' = 4x + 3y - 2$.

363. $y^2 y' + x^2 \sin^3 x = y^3 \operatorname{ctg} x$.

364. $2xy' - y = \sin y'$.

365. $(x^2 y^2 + 1) y + (xy - 1)^2 xy' = 0$.

366. $y \sin x + y' \cos x = 1$.

367. $x \, dy - y \, dx = x\sqrt{x^2 + y^2} \, dx$.

368. $y^2 + x^2 y'^5 = xy (y'^2 + y'^3)$.

369. $y' = \sqrt{2x - y} + 2$.

370. $\left(x - y \cos \dfrac{y}{x}\right) dx + x \cos \dfrac{y}{x} \, dy = 0$.

371. $2\left(x^2 y + \sqrt{1 + x^4 y^2}\right) dx + x^3 \, dy = 0$.

372. $(y' - x\sqrt{y})(x^2 - 1) = xy$.

373. $y'^3 + (y'^2 - 2y) x = 3y' - y$.

374. $(2x + 3y - 1) dx + (4x + 6y - 5) dy = 0$.

375. $(2xy^2 - y) dx + (y^2 + x + y) dy = 0$.

376. $y = y'\sqrt{1 + y'^2}$.   **377.** $y^2 = (xyy' + 1) \ln x$.

378. $4y = x^2 + y'^2$.

379. $2x \, dy + y \, dx + xy^2 (x \, dy + y \, dx) = 0$.

380. $x \, dx + (x^2 \operatorname{ctg} y - 3 \cos y) dy = 0$.

381. $x^2 y'^2 - 2(xy - 2)y' + y^2 = 0.$

382. $xy' + 1 = e^{x-y}.$ **383.** $y' = \operatorname{tg}(y - 2x).$

384. $3x^2 - y = y' \sqrt{x^2 + 1}.$ **385.** $yy' + xy = x^3.$

386. $x(x-1)y' + y^3 = xy.$ **387.** $xy' = 2y + \sqrt{1 + y'^2}.$

388. $(2x + y + 5)y' = 3x + 6.$ **389.** $y' + \operatorname{tg} y = x \sec y.$

390. $y'^4 = 4y(xy - 2y)^2.$ **391.** $y' = \dfrac{y^2 - x}{2y(x+1)}.$

392. $xy' = x^2 e^{-y} + 2.$ **393.** $y' = 3x + \sqrt{y - x^2}.$

394. $x\,dy - 2y\,dx + xy^2(2x\,dy + y\,dx) = 0.$

395. $(x^3 - 2xy^2)\,dx + 3x^2\,y\,dy = x\,dy - y\,dx.$

396. $(yy')^3 = 27x(y^2 - 2x^2).$ **397.** $y' - 8x\sqrt{y} = \dfrac{4xy}{x^2 - 1}.$

398. $[2x - \ln(y + 1)]\,dx - \dfrac{x + y}{y + 1}\,dy = 0.$

399. $xy' = (x^2 + \operatorname{tg} y)\cos^2 y.$ **400.** $x^2(y - xy') = yy'^2.$

401. $y' = \dfrac{3x^2}{x^3 + y + 1}.$ **402.** $y' = \dfrac{(1 + y)^2}{x(y + 1) - x^2}.$

403. $(y - 2xy')^2 = 4yy'^3.$

404. $6x^5 y\,dx + (y^4 \ln y - 3x^6)\,dy = 0.$

405. $y' = \dfrac{1}{2}\sqrt{x} + \sqrt{y}.$ **406.** $2xy' + 1 = y + \dfrac{x^2}{y - 1}.$

407. $yy' + x = \dfrac{1}{2}\left(\dfrac{x^2 + y^2}{x}\right)^2.$ **408.** $y' = \left(\dfrac{3x + y^3 - 1}{y}\right)^2.$

409. $\left(x\sqrt{y^2 + 1} + 1\right)(y^2 + 1)\,dx = xy\,dy.$

410. $(x^2 + y^2 + 1)yy' + (x^2 + y^2 - 1)x = 0.$

411. $y^2(x - 1)\,dx = x(xy + x - 2y)\,dy.$

412. $(xy' - y)^2 = x^2 y^2 - x^4.$

413. $xyy' - x^2\sqrt{y^2 + 1} = (x + 1)(y^2 + 1).$

414. $(x^2 - 1)y' + y^2 - 2xy + 1 = 0.$

415. $y' \operatorname{tg} y + 4x^3 \cos y = 2x.$

416. $(xy' - y)^2 = y'^2 - \dfrac{2yy'}{x} + 1.$

417. $(x + y)(1 - xy)\,dx + (x + 2y)\,dy = 0.$

418. $(3xy + x + y)y\,dx + (4xy + x + 2y)x\,dy = 0.$

419. $(x^2 - 1)\,dx + (x^2 y^2 + x^3 + x)\,dy = 0.$

420. $x(y'^2 + e^{2y}) = -2y'.$

Section 10

EQUATIONS WHICH CAN BE REDUCED TO
EQUATIONS OF LOWER ORDER

If a differential equation does not involve the dependent

variable y explicitly, the order of the equation can be de-

pressed by introducing a new dependent variable. In particular

the equation

$$F(x, y^{(k)}, y^{(k + 1)}, \ldots, y^{(n)}) = 0$$

can be written as an equation of order $n - k$ by introducing

the new variable $y^{(k)} = z$.

If a differential equation does not involve the independent

variable x explicitly, the order can be reduced by rewriting

the equation so that the dependent variable y becomes the

independent variable. Thus if the original equation is

$$F(y, y', y'', \ldots, y^{(n)}) = 0,$$

the equation is rewritten by using the device

$$y' = p(y),$$

where $p(y)$ is an unknown function of y.

<u>Example</u>. Solve the equation $2 y y'' = y'^2 + 1$.

Solution. Since x does not appear explicitly, the device $y' = p(y)$ avails. Thus

$$y'' = \frac{d(y')}{dx} = \frac{dp(y)}{dx} = \frac{dp}{dy} \cdot \frac{dy}{dx} = p'p.$$

From $y' = p$, we obtained $y'' = p p'$, and the equation becomes

$$2 y p p' = p^2 + 1.$$

This equation has lower order than did the original equation. Its solution is $p = \pm\sqrt{Cy - 1}$. From this, $y' = \pm\sqrt{Cy - 1}$. Solving the latter, we obtain

$$4(Cy - 1) = C^2 (x + C_2).$$

If a differential equation is homogeneous in the symbols y, y', y'', ..., i.e. if the equation is unaltered when these symbols are replaced by ky, ky', ky'', $\cdots$, the order of the equation can be reduced by taking $y' = y z$, where z is a new unknown function.

The order of an equation can be reduced if it is homogeneous (in the extended sense) in the variables x, y. This means that the equation is not changed if x is replaced by kx, y by $k^m y$, y' by $k^{m-1} y'$, y'' by $k^{m-2} y''$, etc. The possibility of finding m depends on the possibility of satis-

fying a set of simultaneous conditions which state that the

degree of homogeneity of each term must be the same. Thus, the

equation

$$2 \; x^4 \; y'' - 3 \; y^2 = x^4$$

is homogeneous in the extended sense if the degrees

$$4 + (m - 2), \quad 2m, \quad 4$$

of the three terms can be made the same. This does occur for

$m = 2$.

Having obtained the value of m   (if it exists) one

introduces the new variable $z = z(t)$ by the relations

$$x = e^t, \; y = ze^{mt}$$

In the new equation, the variable t does not appear explicitly.

Thus its order can be reduced.

Thus for example,

$$dy/dx = e^{-t} \; dy/dt, \quad d^2y/dx^2 = e^{-t} \; d/dt \; (e^{-t} \; dy/dt).$$

The order of an equation can be reduced if the equation

can be written so that each member is a polynomial in the

derivative y', or of some function of y'. For example, on

division by $y \, y'$, the equation $y \, y'' = y'^2$ becomes

$$\frac{y''}{y'} = \frac{y'}{y}; \quad (\ln y')' = (\ln y)'; \quad \ln y' = \ln y + \ln C; \quad y' = yC.$$

and the last equation has lower order.

Solve equations 421-450.

421. $x^2 y'' = y'^2$.   **422.** $2xy'y'' = y'^2 - 1$.

423. $y^3 y'' = 1$.   **424.** $y'^2 + 2yy'' = 0$.

425. $y'' = 2yy'$.   **426.** $yy'' + 1 = y'^2$.

427. $y''(e^x + 1) + y' = 0$.   **428.** $y''' = y''^2$.

429. $yy'' = y'^2 - y'^3$.   **430.** $y''' = 2(y'' - 1) \operatorname{ctg} x$.

431. $2yy'' = y^2 + y'^2$.   **432.** $y''^3 + xy'' = 2y'$.

433. $y''^2 + y' = xy''$.   **434.** $y'' + y'^2 = 2e^{-y}$.

435. $x^2 y''' = y''^2$.   **436.** $y''^2 = y'^2 + 1$.

437. $y'' = e^y$.   **438.** $y'' - xy''' + y'''^3 = 0$

439. $2y'(y'' + 2) = xy''^2$.   **440.** $y^4 - y^3 y'' = 1$.

441. $y'^2 = (3y - 2y')y''$.   **442.** $y''(2y' + x) = 1$.

443. $y''^2 - 2y'y''' + 1 = 0$.   **444.** $(1 - x^2)y'' + xy' = 2$.

445. $yy'' - 2yy' \ln y = y'^2$.   **446.** $(y' + 2y)y'' = y'^2$.

447. $xy'' = y' + x \sin \dfrac{y'}{x}$.   **448.** $y'''y'^2 = y''^3$.

449. $yy'' + y = y'^2$.   **450.** $xy'' = y' + x(y'^2 + x^2)$.

Solve equations 451-454. Use integration by parts to reduce any iterated integral to a single integration. For example,

$$\int_0^x \int_0^s \frac{\sin t}{t}\, dt\, ds$$

$$= s \int_0^s \frac{\sin t}{t}\, dt \Bigg]_0^x - \int_0^x \sin s\, ds$$

451. $xy^{IV} = 1$.   **452.** $xy'' = \sin x$.

453. $y''' = 2xv''$ **454.** $xy^{IV} + y''' = e^x$.

Solve equations 455-462 by rewriting so that each member becomes a polynomial in the derivative y' (or something similar).

455. $yy''' = y'y''.$

456. $y'y''' = 2y''^2.$

457. $yy'' = y'(y' + 1).$

458. $5y'''^2 - 3y''y^{IV} = 0.$

459. $yy'' + y'^2 = 1.$

460. $y'' = xy' + y + 1.$

461. $xy'' = 2yy' - y'$

462. $xy'' - y' = x^2yy'.$

In problems 463-480, reduce the order of the given equation by noting that it is homogeneous. Solve the equation.

463. $xyy'' - xy'^2 = yy'.$

464. $yy'' = y'^2 + 15y^2\sqrt{x}.$

465. $(x^2 + 1)(y'^2 - yy'') = xyy'.$

466. $xyy'' + xy'^2 = 2yy'.$

467. $x^2yy'' = (y - xy')^2.$

468. $y'' + \dfrac{y'}{x} + \dfrac{y}{x^2} = \dfrac{y'^2}{y}.$

469. $y(xy'' + y') = xy'^2(1 - x).$

470. $x^2yy'' + y'^2 = 0.$

471. $x^2(y'^2 - 2yy'') = y^2.$

472. $xyy'' = y'(y + y').$

473. $4x^2y^3y'' = x^2 - y^4.$

474. $x^3y'' = (y - xy')(y - xy' - x).$

475. $\dfrac{y^2}{x^2} + y'^2 = 3xy' + \dfrac{2yy'}{x}.$

476. $y'' = \left(2xy - \dfrac{5}{x}\right)y' + 4y^2 - \dfrac{4y}{x^2}.$

477. $x^2(2yy'' - y'^2) = 1 - 2xyy'.$

478. $x^2(yy'' - y'^2) + xyy' = (2xy' - 3y)\sqrt{x^3}.$

479. $x^4(y'^2 - 2yy'') = 4x^3yy' + 1.$

480. $yy' + xyy'' - xy'^2 = x^3.$

In problems 481-500, reduce the order of the given equation to first order.

481. $y''(3 + yy'^2) = y'^4$.   **482.** $y''^2 - y'y''' = \left(\dfrac{y'}{x}\right)^2$.

483. $yy' + 2x^2y'' = xy'^2$.   **484.** $y'^2 + 2xyy'' = 0$.

485. $2xy^2(xy'' + y') + 1 = 0$.   **486.** $x(y'' + y'^2) = y'^2 + y'$.

487. $y^2(y'y''' - 2y''^2) = y'^4$.   **488.** $y(2xy'' + y') = xy'^2 + 1$.

489. $y'' + 2yy'^2 = \left(2x + \dfrac{1}{x}\right)y'$.

490. $y'y''' = y''^2 + y'^2y''$.   **491.** $yy'' = y'^2 + 2xy^2$.

492. $y''^4 = y'^5 - yy'^3y''$.   **493.** $2yy''' = y'$.

494. $y'''y'^2 = 1$.   **495.** $y^2y''' = y'^3$.

496. $x^2yy'' + 1 = (1 - y)xy'$.

497. $y^2(x^3y''' - 2xy' - 3y) = x^3y'(3yy'' - 2y'^2)$.

498. $(y'y''' - 3y''^2)y = y'^5$.

499. $y^2(y'y'' - 2y''^2) = yy'^2y'' + 2y'^4$.

500. $x^2(y^2y''' - y'^3) = 2y^2y' - 3xyy'^2$.

In problems 501-505, find the solution (or solutions) that satisfy the given conditions.

501. $yy'' = 2xy'^2$; $y(2) = 2$, $y'(2) = 0,5$.

502. $2y''' - 3y'^2 = 0$; $y(0) = -3$, $y'(0) = 1$, $y''(0) = -1$.

503. $x^2y'' - 3xy' = \dfrac{6y^2}{x^2} - 4y$; $y(1) = 1$, $y'(1) = 4$.

504. $y''' = 3yy'$; $y(0) = -2$, $y'(0) = 0$, $y''(0) = 4,5$.

505. $y'' \cos y + y'^2 \sin y = y'$; $y(-1) = \dfrac{\pi}{6}$, $y'(-1) = 2$.

506. Find a curve such that the radius of curvature at an arbitrary point is twice the length of the segment of the normal drawn from this point to the axis of abscissas. Consider two cases: (a) the curve is convex to the axis of abscissas; (b) the curve is concave.

507. Find a curve in which the radius of curvature at an arbitrary point is inversely proportional to the cosine of the angle between the tangent line and the axis of abscissas.

508. Find the shape of a uniform inextensible filament fastened at both ends, if the filament supports a load such that the horizontal projection of the load supported by each unit of length is constant (suspension bridge cable). Neglect the weight of the filament. Denote the load of the horizontal projection per unit length by p, and the horizontal component of the tension in the filament by T (a constant).

509. Find the shape of a uniform filament suspended under its own weight (fixed at both ends). Let q be the weight of a unit length of the filament, and set a = q/T, where T is defined in problem 508.

510. Show that the pendulum equation

$$y'' + \sin y = 0$$

has a particular solution $y(x)$ that approaches π for $x \to \infty$.

Section 11

LINEAR EQUATIONS WITH CONSTANT COEFFICIENTS

Every elementary textbook gives several methods for solving these equations. It may be convenient to set $\sin x = \text{Im } e^{xi}$, $\cos 3x = \text{Re } e^{3xi}$, etc.

In the method of undetermined coefficients, one looks for solutions of the form e^{mx}, $e^{mx} \cos px$, $x^r e^{mx} \cos px$.

The method of adjusting the unknown parameters in a function is useful in solving a wide class of problems. For example, the function $y = \exp mx$ satisfies the equation

$$y'(x) = y (x - 1)$$

provided m satisfies the complicated equation $m = \exp(-m)$; thus m must have the value 0.567 if m is real.

Solve problems 511-547.

511. $y'' + y' - 2y = 0.$ **512.** $y'' + 4y' + 3y = 0.$

513. $y'' - 2y' = 0.$ **514.** $2y'' - 5y' + 2y = 0.$

515. $y'' - 4y' + 5y = 0.$ **516.** $y'' + 2y' + 10y = 0.$

517. $y'' + 4y = 0.$ **518.** $y''' - 8y = 0.$

519. $y^{IV} - y = 0.$ **520.** $y^{IV} + 4y = 0.$

521. $y^{VI} + 64y = 0.$ **522.** $y'' - 2y' + y = 0.$

523. $4y'' + 4y' + y = 0.$ **524.** $y^{V} - 6y^{IV} + 9y''' = 0.$

525. $y^{V} - 10y''' + 9y' = 0.$ **526.** $y^{IV} + 2y'' + y = 0.$

527. $y''' - 3y'' + 3y' - y = 0.$ **528.** $y''' - y'' - y' + y = 0.$

529. $y^{IV} - 5y'' + 4y = 0.$ **530.** $y^{V} + 8y''' + 16y' = 0.$

531. $y''' - 3y' + 2y = 0.$ **532.** $y^{IV} + 4y'' + 3y = 0.$

533. $y'' - 2y' - 3y = e^{4x}.$ **534.** $y'' + y = 4xe^{x}.$

535. $y'' - y = 2e^{x} - x^{2}.$ **536.** $y'' + y' - 2y = 3xe^{x}.$

537. $y'' - 3y' + 2y = \sin x.$ **538.** $y'' + y = 4 \sin x.$

539. $y'' - 5y' + 4y = 4x^{2}e^{2x}.$ **540.** $y'' - 3y' + 2y = x \cos x.$

541. $y'' + 3y' - 4y = e^{-4x} + xe^{-x}.$

542. $y'' + 2y' - 3y = x^{2}e^{x}.$

543. $y'' - 4y' + 8y = e^{2x} + \sin 2x.$

544. $y'' - 9y = e^{3x} \cos x.$ **545.** $y'' - 2y' + y = 6xe^{x}.$

546. $y'' + y = x \sin x.$ **547.** $y'' + 4y' + 4y = xe^{2x}.$

548. $y'' - 5y' = 3x^{2} + \sin 5x.$

In problems 549-574, show the form of solution to look for

(with undetermined coefficients).

549. $y'' - 2y' + 2y = e^{x} + x \cos x.$

550. $y'' + 6y' + 10y = 3xe^{-3x} - 2e^{3x} \cos x.$

551. $y'' - 8y' + 20y = 5xe^{4x} \sin 2x.$

552. $y'' + 7y' + 10y = xe^{-2x} \cos 5x.$

553. $y'' - 2y' + 5y = 2xe^{x} + e^{x} \sin 2x.$

554. $y'' - 2y' + y = 2xe^{x} + e^{x} \sin 2x.$

555. $y'' - 8y' + 17y = e^{4x}(x^{2} - 3x \sin x).$

556. $y''' + y' = \sin x + x \cos x.$

557. $y''' - 2y'' + 4y' - 8y = e^{2x} \sin 2x + 2x^{2}.$

558. $y'' - 6y' + 8y = 5xe^{2x} + 2e^{4x} \sin x.$

559. $y'' + 2y' + y = x(e^{-x} - \cos x).$

560. $y''' - y'' - y' + y = 3e^{x} + 5x \sin x.$

561. $y'' - 6y' + 13y = x^{2}e^{3x} - 3 \cos 2x.$

562. $y'' - 9y = e^{-3x}(x^{2} + \sin 3x).$

563. $y^{IV} + y'' = 7x - 3 \cos x.$ **564.** $y'' + 4y = \cos x \cdot \cos 3x.$

565. $y''' - 4y'' + 3y' = x^{2} + xe^{2x}.$

566. $y'' - 4y' + 5y = e^{2x} \sin^{2} x.$

567. $y'' + 3y' + 2y = e^{-x} \cos^{2} x.$

568. $y'' - 2y' + 2y = (x + e^{x}) \sin x.$

569. $y^{IV} + 5y'' + 4y = \sin x \cdot \cos 2x.$

570. $y'' - 3y' + 2y = 2^{x}.$ **571.** $y'' - y = 4 \operatorname{sh} x.$

572. $y'' + 4y' + 3y = \operatorname{ch} x.$ **573.** $y'' + 4y = \operatorname{sh} x \cdot \sin 2x.$

574. $y'' + 2y' + 2y = \operatorname{ch} x \cdot \sin x.$

Solve equations 575-580 by the method of variation of constants. That is, set up a general solution of the reduced equation, and replace the constants c_1, c_2 by appropriate functions of x.

575. $y'' - 2y' + y = \dfrac{e^x}{x}$.   **576.** $y'' + 3y' + 2y = \dfrac{1}{e^x + 1}$.

577. $y'' + y = \dfrac{1}{\sin x}$.   **578.** $y'' + 4y = 2 \operatorname{tg} x$.

579. $y'' + 2y' + y = 3e^{-x} \sqrt{x + 1}$.

580*. $x^3 (y'' - y) = x^2 - 2$.

Find the solutions of 581-588 that satisfy the given conditions.

581. $y''' - y' = 0$; $y(0) = 3$, $y'(0) = -1$, $y''(0) = 1$.

582. $y'' - 2y' + y = 0$; $y(2) = 1$, $y'(2) = -2$.

583. $y'' + y = 4e^x$; $y(0) = 4$, $y'(0) = -3$.

584. $y'' - 2y' = 2e^x$; $y(1) = -1$, $y'(1) = 0$.

585. $y'' - y = 2x$; $y(0) = 0$, $y(1) = -1$.

586. $y'' + y = 1$; $y(0) = 0$, $y\left(\dfrac{\pi}{2}\right) = 0$.

587. $y'' + y = 1$; $y(0) = 0$, $y(\pi) = 0$.

588. $y'' + y = 2x - \pi$; $y(0) = 0$. $y(\pi) = 0$.

In problems 589-600, solve the Euler equation. Solutions will have the form x^m, $(\ln x)^r x^m$, $x^m \cos(p \ln x)$, $(\ln x)^r x^m \cos(p \ln x)$. Or set $x = e^t$ and reduce the equation to the preceding form.

589. $x^2y'' - 4xy' + 6y = 0.$ **590.** $x^2y'' - xy' - 3y = 0.$

591. $x^3y''' + xy' - y = 0.$ **592.** $x^2y''' = 2y'.$

593. $x^2y'' - xy' + y = 8x^3.$ **594.** $x^2y'' + xy' + 4y = 10x.$

595. $x^3y'' - 2xy = 6 \ln x.$ **596.** $x^2y'' - 3xy' + 5y = 3x^2.$

597. $x^2y'' - 6y = 5x^3 + 8x^2.$ **598.** $x^2y'' - 2y = \sin \ln x.$

599. $(x - 2)^2 y'' - 3(x - 2) y' + 4y = x.$

600. $(2x + 3)^3 y''' + 3(2x + 3) y' - 6y = 0.$

LINEAR EQUATIONS WITH VARIABLE COEFFICIENTS

Problems 601-655 are to be solved by the use of methods described in any book on the general theory of linear differential equations. Each problem has either a hint or a reference to the literature.

When a particular solutions y_1 is known for a linear equation of order n in reduced form (also called homogeneous form, or having 0 in the right member) and the equation of lower order is again a linear equation. The method is to set $y = zy_1$ and then reduce the order by making the substitution $u = z'$.

In case the given equation has second order

$$a_o(x) \, y'' + a_1(x) \, y' + a_2(x) \, y = 0,$$

a convenient method for finding a general solution when one solution y_1 is known is to use the formula of Liouville-Ostrogradskii:

$$\begin{vmatrix} y_1 & y_2 \\ y_1' & y_2' \end{vmatrix} = Ce^{-\int p(x)dx}, \quad p(x) = \frac{a_1(x)}{a_o(x)},$$

where y_1 and y_2 are any two solutions of a given equation.

<u>Example</u>. We start with the known solution $y_1 = x$ of the equation

$$(x^2 + 1) y'' - 2xy' + 2y = 0. \tag{1}$$

The Liouville-Ostrogradskii formula gives

$$\begin{vmatrix} y_1 & y_2 \\ y_1' & y_2' \end{vmatrix} = Ce^{-\int (\frac{-2x}{x^2 + 1}) \, dx}; \quad y_1 y_2' - y_1' y_2 = C(x^2 + 1).$$

Since the function y_1 is known, this gives a linear first order equation for y_2. This can be solved in several ways; for instance, if we divide the equation by y_1^2, the left member becomes the derivative of the quotient y_2/y_1:

$$\left(\frac{y_2}{y_1}\right)' = \frac{y_1 y_2' - y_1' y_2}{y_1^2} = \frac{C(x^2 + 1)}{y_1^2}.$$

Since $y_1 = x$, we obtain

$$\frac{y_2}{y_1} = \int C \cdot \frac{x^2 + 1}{x^2} \, dx + C_2 = C\left(x - \frac{1}{x}\right) + C_2;$$

$$y = C(x^2 - 1) + C_2 x.$$

Thus we have the general solution of equation (1).

There is no general method for finding even a particular solution of a linear second order differential equation, but in some cases we can find the solution of a specified form by the method of undetermined coefficients.

Example. Let us try to find a solution of the equation

$$(1 - 2x^2) \, y'' + 2y' + 4y = 0 \, , \tag{2}$$

which is a rational polynomial, if there is one.

The first step is to find what degree the polynomial must have. If we put in the term x^n of highest degree in equation (2) and compute the terms of highest degree throughout, we find that only the term involving the second derivative and the terms involving y will contain the term of highest degree:

$$- 2x^2 \cdot n(n-1) \, x^{n-2} + \ldots + 4x^n + \ldots = 0.$$

Since the coefficient of x^n must be 0, we see that the relation $- 2n \, (n - 1) + 4 = 0$; $n^2 - n - 2 = 0$ must hold. There are two solutions, $n = 2$, $n = -1$, the latter of which is not useful since a polynomial cannot have negative degree. Thus the polynomial has second degree and takes the form $y = x^2 + ax + b$. If we insert this in equation (2), we obtain the relation $(4a + 4) \, x + 2 + 2a + 4b = 0$. Therefore, $4a + 4 = 0$, $2 + 2a + 4b = 0$. This gives $a = -1$, $b = 0$. Thus the polynomial $y = x^2 - x$ is a particular solution.

In problems 601-622 determine whether the given functions are linearly independent. In each case the functions of the set are to be taken in the region in which all are defined.

601. $x + 2$, $x - 2$.   **602.** $6x + 9$, $8x + 12$.
603. $\sin x$, $\cos x$.   **604.** 1, x, x^2.
605. $4 - x$, $2x + 3$, $6x + 8$.
606. $x^2 + 2x$, $3x^2 - 1$, $x + 4$.
607. $x^2 - x + 3$, $2x^2 + x$, $2x - 4$.
608. e^x, e^{2x}, e^{3x}.   **609.** x, e^x, xe^x.
610. 1, $\sin^2 x$, $\cos 2x$.   **611.** $\operatorname{sh} x$, $\operatorname{ch} x$, $2 + e$.
612. $\ln(x^2)$, $\ln 3x$, 7.   **613.** x, 0, e^x.
614. $\operatorname{sh} x$, $\operatorname{ch} x$, $2e^x - 1$, $3e^x + 5$.
615. 2^x, 3^x, 6^x.   **616.** $\sin x$, $\cos x$, $\sin 2x$.
617. $\sin x$, $\sin(x + 2)$, $\cos(x - 5)$.
618. $\sqrt{x}$, $\sqrt{x+1}$, $\sqrt{x+2}$.
619. $\operatorname{arctg} x$, $\operatorname{arcctg} x$, 1.   **620.** x^2, $x|x|$.
621. x, $|x|$, $2x + \sqrt{4x^2}$.   **622.** x, x^3, $|x^3|$.

623.[*] Let the functions $y_1(x)$ and $y_2(x)$ be linearly independent on an interval (a,b). If these functions are differentiable and if the determinant $y_1 y_2' - y_2 y_1'$ has the value 0, show that there is a point x_o on the given interval for which the relations $y_1(x_o) = y_2(x_o) = y_1'(x_o) = y_2'(x_o) = 0$ hold.

In each of the problems 624-630 find a linear homogeneous differential equation which has the given particular solutions and as low an order as possible.

624. 1, $\cos x$.   **625.** x, e^x.
626. $3x$, $x - 2$, $e^x + 1$.
627. $x^2 - 3x$, $2x^2 + 9$, $2x + 3$.
628. e^x, $\operatorname{sh} x$, $\operatorname{ch} x$.   **629.** x, x^2, e^x.
630. x, x^3, $|x^3|$.

66

In problems 631-651 find the general solution of the given equation by starting with a particular solution. If no particular solution is given try to find a particular solution of the form $y_1 = e^{ax}$ or a rational polynomial $y = x^n + ax^{n-1} + bx^{n-2} + \cdots$

631. $(2x+1)y'' + 4xy' - 4y = 0.$

632. $x^2(x+1)y'' - 2y = 0;$ $y_1 = 1 + \dfrac{1}{x}.$

633. $xy'' - (2x+1)y' + (x+1)y = 0.$

634. $xy'' + 2y' - xy = 0;$ $y_1 = \dfrac{e^x}{x}.$

635. $y'' - 2(1 + \text{tg}^2 x)y = 0;$ $y_1 = \text{tg}\, x.$

636. $x(x-1)y'' - xy' + y = 0.$

637. $(e^x+1)y'' - 2y' - e^x y = 0;$ $y_1 = e^x - 1.$

638. $x^2 y'' \ln x - xy' + y = 0.$

639. $y'' - y'\, \text{tg}\, x + 2y = 0;$ $y_1 = \sin x.$

640. $(x^2+1)y'' + 5xy' + 4y = 0;$ $y_1 = \dfrac{x}{(x^2+1)^{\frac{3}{2}}}.$

641. $xy'' - (x+1)y' - 2(x-1)y = 0.$

642. $y'' + 4xy' + (4x^2+2)y = 0;$ $y_1 = e^{ax^2}.$

643. $xy'' - (2x+1)y' + 2y = 0.$

644. $x(2x+1)y'' + 2(x+1)y' - 2y = 0.$

645. $x(x+4)y'' - (2x+4)y' + 2y = 0.$

646. $x(x^2+6)y'' - 4(x^2+3)y' + 6xy = 0.$

647. $(x^2+1)y'' - 2y = 0.$

648. $2x(x+2)y'' + (2-x)y' + y = 0.$

649. $xy''' - y'' - xy' + y = 0;$ $y_1 = x,\ y_2 = e^x.$

650. $x^2(2x-1)y''' + (4x-3)xy'' - 2xy' + 2y = 0;$

$$y_1 = x,\ y_2 = \frac{1}{x}.$$

651. $(x^2-2x+3)y''' - (x^2+1)y'' + 2xy' - 2y = 0;$

$$y_1 = x,\ y_2 = e^x.$$

Find a general solution of the linear non-homogeneous equations given, knowing that a particular solution of the corresponding homogeneous (reduced) equation is a polynomial.

Knowing two particular solutions of the linear non-homogeneous second order given, find the general solution.

In problems 656-660 make a substitution $y = a(x)z$ to obtain a differential equation with the term involving the first derivative absent.

656. $x^2y'' - 2xy' + (x^2 + 2)y = 0.$
657. $x^2y'' - 4xy' + (6 - x^2)y = 0.$
658. $(1 + x^2)y'' + 4xy' + 2y = 0.$
659. $x^2y'' + 2x^2y' + (x^2 - 2)y = 0.$
660. $xy'' + y' + xy = 0.$

In problems 661-665 change the independent variable by the formula $t = \phi(x)$ so that the term involving the first derivative is absent. Note that the function $\phi(x)$ is not uniquely defined by the requirements of the problem because if $\phi(x)$ solves the problem, so does $a\phi(x) + b$ for arbitrary constant a,b.

661. $xy'' - y' - 4x^3y = 0.$
662. $(1 + x^2)y'' + xy' + y = 0.$
663. $x^2(1 - x^2)y'' + 2(x - x^3)y' - 2y = 0.$
664. $y'' - y' + e^{1x}y = 0.$
665. $2xy'' + y' + xy = 0.$

666. Find the distance between two adjacent zeros of a solution ($\neq 0$) of the equation $y'' + my = 0$, where m is a positive constant. How many zeros can be contained on the interval $a \leq x \leq b$?

In problems 667-670 use the Sturm-Liouville theory to find the largest and smallest possible numbers of zeros of a solution of the corresponding equation, providing the solution is not identically 0, on the given interval.

667. $y'' + 2xy = 0$, $20 \leqslant x \leqslant 45$.
668. $xy'' + y = 0$, $25 \leqslant x \leqslant 100$.
669. $y'' - 2xy' + (x+1)^2 y = 0$, $4 \leqslant x \leqslant 19$.
670. $y'' - 2e^x y' + e^{2x} y = 0$, $2 \leqslant x \leqslant 6$.

671.* Solve the corresponding problem for the equation $y'' + xy = 0$ on the segment $-25 \leq x \leq 25$.

672. Let x_1, x_2, ... be the distinct zeros, arranged in ascending order, of a solution of the equation $y'' + q(x)y = 0$, where $q(x) > 0$. Suppose the function $q(x)$ is continuous and increasing for $A < x < \infty$. Show that the relation $x_{n+1} - x_n < x_n - x_{n-1}$ holds; that is that the distance between successive zeros decreases.

673. In the preceding problem let c be the finite limit or ∞, of the function $q(x)$ as $x \to \infty$. Show that the relation

$$\lim_{n \to \infty} (x_{n+1} - x_n) = \frac{\pi}{\sqrt{c}}$$

holds.

674.* Let the functions $y(x)$ and $z(x)$ satisfy the equations $y'' + q(x) y = 0$ and $z'' + Q(x) z = 0$, let $y(x)$ be positive on the segment (x_1, x_2) and let $z(x)$ be positive on the segment (x_1, x_2^*), and 0 at each end point. Suppose $Q(x) > q(x) > 0$. Show that if the relation $z'(x_1) \not{=} y'(x_1)$ holds, then for all x on $x_1 < x \leq x_2^*$, the relation $z(x) < y(x)$ holds.

675.* In connection with problem 672, set

$$b_n = \max_{x_n \leq x \leq x_{n+1}} y(x) .$$

Show that $b_1 > b_2 > b_3 > \cdots$

676.* In problem 673, suppose the limit c is finite. Show that $b_n \to B > 0$, for $n \to \infty$, where b_n is defined in problem 675.

677. By a change $t = \phi(x)$ of the independent variable, rewrite the equation $d^2 y/dx^2 \pm y/(\psi(x))^4 = 0$ in the form $d^2 y/dt^2 + b(t)\, dy/dt \pm y = 0$, and then remove the term involving the first derivative by making the substitution $y = a(t)\, u$.

Note: This transformation is due to Liouville. This transformation converts the equation $y'' + q(x)\, y = 0$ into an equation of the same form, but with coefficients that are nearly constant, that is that change very little on the interval (t_o, ∞). In that way it may be possible to obtain information about the behavior of the solution in the limit $x \to \infty$.

678.[*] Suppose the relation $|f(t)| \le c/t^{1+d}$, for some positive $d > 0$. Show that there are two solutions $u_1(t)$, $u(t)$ of the equation $u'' + (1 + f(t))\, u = 0$ with the properties $u_1(t)$ $\cos t + 0\,(t^{-d})$, $u_2(t) = \sin t + 0\,(t^{-d})$, $t \to \infty$.

Hint. Transpose the term $f(t)\, u$ to the right member, denote this term by $F(t)$ and use the method of variation of constants. The integrals involved in the solution by this method are to have upper limit $+\infty$. Then use the method of successive approximations, taking the initial approximation to be $u = \cos t$ $\left[u = \sin t. \right]$

679.* Under the same hypothesis as in 678, consider the equation $u'' - (1 - f(t)) u = 0$ and obtain the relations

$$u_1(t) = e^t(1 + O(t^{-d})), \quad u_2(t) = e^{-t}(1 + O(t^{-d})).$$

In problems 680-688 use the method of Liouville, explained after problem 677, to find the asymptotic behavior of the solutions of the given equation $x \to \infty$. Make use of the assertions of problems 678, 679.

680. $y'' + x^2 y = 0.$ **681.** $y'' + e^{2x} y = 0.$
682. $xy'' - y = 0.$ **683.** $y'' - xy = 0.$
684. $xy'' + 2y' + y = 0.$
685. $y'' - 2(x - 1) y' + x^2 y = 0.$
686*. $y'' + (x^4 + 1) y = 0.$ **687*.** $(x^2 + 1) y'' - y = 0.$
688*. $x^2 y'' + y \ln^2 x = 0.$

In problems 689-690 use Liouville's transformation twice to get two terms in the asymptotic behavior of the solutions of the given equations.

689. $y'' - 4x^2 y = 0.$ **690.** $xy'' + y = 0.$

Section 13

SERIES DEVELOPMENT OF SOLUTIONS OF EQUATIONS

A function $f(x,y)$ that is analytic in the neighborhood
of the point (x_o, y_o) can be expanded in a series of powers of
$(x-x_o)$ and $(y-y_o)$. Moreover, the equation $y' = f(x,y)$ has a
solution that takes the initial value $y(x_o) = y_o$ and is an
analytic function of x. This is shown in standard works on the
existence of solutions of differential equations. A corresponding theorem holds for an equation of the form
$y^{(n)} = f(x,y,y',\ldots y^{(n-1)})$ with initial conditions
$y(x_o) = y_o,\; y'(x_o) = y_o,\ldots,y^{(n-1)}(x_o) = y_o^{(n-1)}$.

Example. To find a series solution for the equation
$y'' = xy^2 - y'$ having initial values $y(0) = 2,\; y'(0) = 1$.

Suppose the series starts as follows ₊

$$y = a_o + a_1 x + a_2 x^2 + \ldots = 2 + x + a_2 x^2 + a_3 x^3 + \ldots, \quad (1)$$

the first two terms having been obtained from the initial values.
If we substitute this series in the differential equation, we
obtain $2a_2 + 6a_3 x + 12a_4 x^2 + \ldots$
$$= x(2 + x + a_2 x^2 + \ldots)^2 - 1 - 2a_2 x - 3a_3 x^2 - \ldots$$

We must now expand the indicated square in the right member and compare coefficients of corresponding powers of x on both sides. This gives $2a_2 = -1$, $6a_3 = 4 - 2a_2$, $12a_4 = 4 - 3a_3$, ... from which we obtain $a_2 = -1/2$, $a_3 = 5/6$, $a_4 = 1/8$, ... Therefore, $y = 2 + x - (1/2)x^2 + (5/6)x^3 + (1/8)x^4 + \ldots$

If we have to solve an equation

$$p_o(x)\, y^{(n)} + p_1(x)\, y^{(n-1)} + \ldots + p_n(x)\, y = 0, \qquad (2)$$

and it happens that $p_o(x_o) = 0$, that is that the coefficient of the highest derivative is zero at the point x_o, there may be no solution in series, but there may be a solution of the form

$$a_o(x - x_o)^r + a_1(x - x_o)^{r+1} + a_2(x - x_o)^{r+2} + \ldots, (3)$$

where the number r is not necessarily a positive integer. The value of r, if there is a suitable one, is found by substituting this series into equation (2) and trying to compare the lowest powers of $(x - x_o)$. Having determined r, one then proceeds to determine the coefficients a_i.

In each of the problems 691-697 find a solution in series which satisfies the given conditions. Compute the first few terms of the series up to the term involving x^4.

691. $y' = y^2 - x;\ y(0) = 1.$ **692.** $y' = x + \dfrac{1}{y};\ y(0) = 1.$

693. $y' = y + xe^y;\quad y(0) = 0.$

694. $y' = 2x + \cos y;\quad y(0) = 0.$

695. $y' = x^2 + y^3;\quad y(1) = 1.$

696. $y'' = xy' - y^2;\quad y(0) = 1,\ y'(0) = 2.$

697. $y'' = y'^2 + xy;\quad y(0) = 4,\ y'(0) = -2.$

698.* Use the relation $- x < y' < 1 + y^2$, valid for

$-1 < x < 1$, to give a lower estimate for the radius of conver-

gence of a power series which represents the solution of the

equation $y' = y^2 - x$ with initial value $y(0) = 1$.

699.* Calculate the accuracy with which a solution of the

equation $y' = e^y - x^2 y$, with initial condition $y(0) = 0$, is

obtained by the use of a series solution going as far as the

term involving x^4, if values of x are restricted to have

absolute value less than 0.2.

In problems 700-709 find the linearly independent solutions

of each of the given equations in the form of series. When pos-

sible, sum the series and express the solutions in terms of

elementary functions.

700. $y'' - x^2 y = 0.$ **701.** $y'' - xy' - 2y = 0.$

702. $(1 - x^2) y'' - 4xy' - 2y = 0.$

703. $(x^2 + 1) y'' + 5xy' + 3y = 0.$

704. $(1 - x) y'' - 2y' + y = 0.$

705. $(x^2 - x + 1) y'' + (4x - 2) y' + 2y = 0.$

706. $y'' - xy' + xy = 0.$ **707.** $y'' + y \sin x = 0.$

708. $xy'' + y \ln (1 - x) = 0.$

709. $y''' - xy'' + (x - 2) y' + y = 0.$

In problems 710-716 find those solutions of the given equations which can be expressed as power series or a generalized power series, Equation (3) above.

710. $xy'' + 2y' + xy = 0$.
711. $2x^2 y'' + (3x - 2x^2) y' - (x + 1) y = 0$.
712. $9x^2 y'' - (x^2 - 2) y = 0$.
713. $x^2 y'' - x^2 y' + (x - 2) y = 0$.
714. $x^2 y'' + 2xy' - (x^2 + 2x + 2) y = 0$.
715. $xy'' - xy' - y = 0$. **716.** $xy'' + y' - xv = 0$.

717. For the solution of problem 716 which is independent of the series solutions found above, find the behavior of the solution, $x \to 0$, to the term of order less than x^5.

In problems 718-720 determine whether solutions exist in the form of power series or generalized power series, see Equation (3) above.

718. $x^2 y'' + xy' - (x + 2) y = 0$.
719. $x^2 y'' + xy' + (1 - x) y = 0$.
720. $x^2 y' + (x - 1) y = -1$.

In problems 721-722 find periodic solutions of the equations in the form of trigonometric series

$$a_o + a_1 \cos x + b_1 \sin x + a_2 \cos 2x + b_2 \sin 2x \ldots$$

Hint. In problem 722 express the right member as a Fourier series of the form $\Sigma 2^{-n} \sin nx$.

721. $y'' + y' + y = |\sin x|$.

722. $y''' - y' - y = \dfrac{2 \sin x}{5 - 4 \cos x}$.

In problems 723-725 develop 2-3 terms of the solution as a series of powers of the parameter μ . Note that the fact that such a series in powers of μ exists follows from the assertion that the solution is an analytic function of μ , as shown in standard works on differential equations. See Sec. 18.

723. $y' = 4\mu(x + 1) - y^2$; $\quad y(0) = 1$.

724. $y' = \dfrac{2}{y} - 5\mu x$; $\qquad y(1) = 2$.

725. $xy' = \mu x^2 + \ln y$; $\qquad y(1) = 1$.

Section 14

LINEAR SYSTEMS WITH CONSTANT COEFFICIENTS

A linear system with constant coefficients can be solved
either by eliminating the unknowns (this process requires
successive differentiations, and requires that the constants
of the solutions so obtained be evaluated by substituting in
the original equations) or can be solved by studying the characteristic equation in determinantal form. A non-homogeneous
system of equations can be solved by obtaining particular
solutions by the method of variation of constants. If the
right members have special forms (polynomials in x, exponen-
tials, sines, cosines, or sums of terms of this form) particular
solutions can always be found by the method of undetermined coefficients as in the case of single equations of corresponding
types.

In problems 726-752 solve the given system of equations.
The symbol $\dot{x}$ is used for dx/dt, $\dot{y}$ for dy/dt, $\dot{z}$ for dz/dt.
The roots of the characteristic equation are indicated in some
problems.

726. $\begin{cases} \dot{x} = 2x + y, \\ \dot{y} = 3x + 4y. \end{cases}$

727. $\begin{cases} \dot{x} = x - y, \\ \dot{y} = y - 4x. \end{cases}$

728. $\begin{cases} \dot{x} + x - 8y = 0, \\ \dot{y} - x - y = 0. \end{cases}$

729. $\begin{cases} \dot{x} = x + y, \\ \dot{y} = 3y - 2x. \end{cases}$

730. $\begin{cases} \dot{x} = x - 3y, \\ \dot{y} = 3x + y. \end{cases}$

731. $\begin{cases} \dot{x} + x + 5y = 0. \\ \dot{y} - x - y = 0. \end{cases}$

732. $\begin{cases} \dot{x} = 2x + y, \\ \dot{y} = 4y - x. \end{cases}$

733. $\begin{cases} \dot{x} = 3x - y, \\ \dot{y} = 4x - y. \end{cases}$

734. $\begin{cases} \dot{x} = 2y - 3x, \\ \dot{y} = y - 2x. \end{cases}$

735. $\begin{cases} \dot{x} - 5x - 3y = 0, \\ \dot{y} + 3x + y = 0. \end{cases}$

736. $\begin{cases} \dot{x} = x + z - y, \\ \dot{y} = x + y - z, \\ \dot{z} = 2x - y \end{cases}$
$(\lambda_1 = 1, \ \lambda_2 = 2, \ \lambda_3 = -1).$

737. $\begin{cases} \dot{x} = x - 2y - z, \\ \dot{y} = y - x + z, \\ \dot{z} = x - z \end{cases}$
$(\lambda_1 = 0, \ \lambda_2 = 2, \ \lambda_3 = -1).$

738. $\begin{cases} \dot{x} = 2x - y + z, \\ \dot{y} = x + 2y - z, \\ \dot{z} = x - y + 2z \end{cases}$
$(\lambda_1 = 1, \ \lambda_2 = 2, \ \lambda_3 = 3).$

739. $\begin{cases} \dot{x} = 3x - y + z, \\ \dot{y} = x + y + z, \\ \dot{z} = 4x - y + 4z \end{cases}$
$(\lambda_1 = 1, \ \lambda_2 = 2, \ \lambda_3 = 5).$

740. $\begin{cases} \dot{x} = 4y - 2z - 3x, \\ \dot{y} = z + x, \\ \dot{z} = 6x - 6y + 5z \end{cases}$
$(\lambda_1 = 1, \ \lambda_2 = 2, \ \lambda_3 = -1).$

741. $\begin{cases} \dot{x} = x - y - z, \\ \dot{y} = x + y, \\ \dot{z} = 3x + z \end{cases}$
$(\lambda_1 = 1, \ \lambda_{2,3} = 1 \pm 2i).$

742. $\begin{cases} \dot{x} = 2x + y, \\ \dot{y} = x + 3y - z, \\ \dot{z} = 2y + 3z - x \end{cases}$
$(\lambda_1 = 2, \ \lambda_{2,3} = 3 \pm i).$

743. $\begin{cases} \dot{x} = 2x + 2z - y, \\ \dot{y} = x + 2z, \\ \dot{z} = y - 2x - z \end{cases}$
$(\lambda_1 = 1, \ \lambda_{2,3} = \pm i).$

744. $\begin{cases} \dot{x} = 4x - y - z, \\ \dot{y} = x + 2y - z, \\ \dot{z} = x - y + 2z \end{cases}$
$(\lambda_1 = 2, \ \lambda_2 = \lambda_3 = 3).$

745. $\begin{cases} \dot{x} = 2x - y - z, \\ \dot{y} = 3x - 2y - 3z, \\ \dot{z} = 2z - x + y \end{cases}$
$(\lambda_1 = 0, \ \lambda_2 = \lambda_3 = 1).$

746. $\begin{cases} \dot{x} = y - 2x - 2z, \\ \dot{y} = x - 2y + 2z, \\ \dot{z} = 3x - 3y + 5z \end{cases}$
$(\lambda_1 = 3, \ \lambda_2 = \lambda_3 = -1).$

747. $\begin{cases} \dot{x} = 3x - 2y - z, \\ \dot{y} = 3x - 4y - 3z, \\ \dot{z} = 2x - 4y \end{cases}$
$(\lambda_1 = \lambda_2 = 2, \ \lambda_3 = -5).$

748. $\begin{cases} \dot{x} = x - y + z, \\ \dot{y} = x + y - z, \\ \dot{z} = 2z - y \end{cases}$
$(\lambda_1 = \lambda_2 = 1, \ \lambda_3 = 2).$

749. $\begin{cases} \dot{x} = y - 2z - x, \\ \dot{y} = 4x + y, \\ \dot{z} = 2x + y - z \end{cases}$
$(\lambda_1 = 1, \ \lambda_2 = \lambda_3 = -1).$

750. $\begin{cases} \dot{x} = 2x + y, \\ \dot{y} = 2y + 4z, \\ \dot{z} = x - z \end{cases}$
$(\lambda_1 = \lambda_2 = 0, \ \lambda_3 = 3).$

751. $\begin{cases} \dot{x} = 2x - y - z, \\ \dot{y} = 2x - y - 2z, \\ \dot{z} = 2z - x + y \end{cases}$
$(\lambda_1 = \lambda_2 = \lambda_3 = 1).$

752. $\begin{cases} \dot{x} = 4x - y, \\ \dot{y} = 3x + y - z, \\ \dot{z} = x + z \end{cases}$
$(\lambda_1 = \lambda_2 = \lambda_3 = 2).$

In problems 753-765 solve the systems without necessarily

reducing them to normal forms.

753. $\begin{cases} \ddot{x} = 2x - 3y, \\ \ddot{y} = x - 2y. \end{cases}$

754. $\begin{cases} \ddot{x} = 3x + 4y, \\ \ddot{y} = -x - y. \end{cases}$

755. $\begin{cases} \ddot{x} = 2y, \\ \ddot{y} = -2x. \end{cases}$

756. $\begin{cases} \ddot{x} = 3x - y - z, \\ \ddot{y} = -x + 3y - z, \\ \ddot{z} = -x - y + 3z. \end{cases}$

757. $\begin{cases} 2\ddot{x} - 5\ddot{y} = 4y - x, \\ 3\ddot{x} - 4\ddot{y} = 2x - y. \end{cases}$

758. $\begin{cases} \ddot{x} + \dot{x} + \dot{y} - 2y = 0, \\ \dot{x} - \dot{y} + x = 0. \end{cases}$

759. $\begin{cases} \ddot{x} - 2\ddot{y} + \dot{y} + x - 3y = 0, \\ 4\ddot{y} - 2\ddot{x} - \dot{x} - 2x + 5y = 0. \end{cases}$

760. $\begin{cases} \ddot{x} - x + 2\ddot{y} - 2y = 0, \\ \dot{x} - x + \dot{y} + y = 0. \end{cases}$

761. $\begin{cases} \ddot{x} - 2\dot{y} + 2x = 0, \\ 3\dot{x} + \ddot{y} - 8y = 0. \end{cases}$

762. $\begin{cases} \ddot{x} + 3\dot{y} - x = 0, \\ \dot{x} + 3\dot{y} - 2y = 0. \end{cases}$

763. $\begin{cases} \ddot{x} + 5\dot{x} + 2\dot{y} + y = 0, \\ 3\ddot{x} + 5x + \dot{y} + 3y = 0. \end{cases}$

764. $\begin{cases} \ddot{x} + 4\dot{x} - 2x - 2\dot{y} - y = 0, \\ \ddot{x} - 4\dot{x} - \ddot{y} + 2\dot{y} + 2y = 0. \end{cases}$

765. $\begin{cases} 2\ddot{x} + 2\dot{x} + x + 3\ddot{y} + \dot{y} + y = 0, \\ \ddot{x} + 4\dot{x} - x + 3\ddot{y} + 2\dot{y} - y = 0. \end{cases}$

In problems 766-785 solve the non-linear non-homogeneous systems.

766. $\begin{cases} \dot{x} = y + 2e^t, \\ \dot{y} = x + t^2. \end{cases}$ **767.** $\begin{cases} \dot{x} = y - 5\cos t, \\ \dot{y} = 2x + y. \end{cases}$

768. $\begin{cases} \dot{x} = 3x + 2y + 4e^{5t}, \\ \dot{y} = x + 2y. \end{cases}$ **769.** $\begin{cases} \dot{x} = 3x - 4y + e^{-2t}, \\ \dot{y} = x - 2y - 3e^{-2t}. \end{cases}$

770. $\begin{cases} \dot{x} = 4x + y - e^{2t}, \\ \dot{y} = y - 2x. \end{cases}$ **771.** $\begin{cases} \dot{x} = 2y - x + 1, \\ \dot{y} = 3y - 2x. \end{cases}$

772. $\begin{cases} \dot{x} = 5x - 3y + 2e^{3t}, \\ \dot{y} = x + y + 5e^{-t}. \end{cases}$ **773.** $\begin{cases} \dot{x} = x + y + 1 + e^t, \\ \dot{y} = 3x - y. \end{cases}$

774. $\begin{cases} \dot{x} = x + 2y, \\ \dot{y} = x - 5\sin t. \end{cases}$ **775.** $\begin{cases} \dot{x} = 2x - 4y, \\ \dot{y} = x - 3y + 3e^t. \end{cases}$

776. $\begin{cases} \dot{x} = 2x - y, \\ \dot{y} = y - 2x + 18t. \end{cases}$ **777.** $\begin{cases} \dot{x} = x + 2y + 16te^t, \\ \dot{y} = 2x - 2y. \end{cases}$

778. $\begin{cases} \dot{x} = 2x + 4y - 8, \\ \dot{y} = 3x + 6y. \end{cases}$ **779.** $\begin{cases} \dot{x} = 2x - 3y, \\ \dot{y} = x - 2y + 2\sin t. \end{cases}$

780. $\begin{cases} \dot{x} = 2x + 3y + 5t, \\ \dot{y} = 3x + 2y + 8e^t. \end{cases}$ **781.** $\begin{cases} \dot{x} = 2x - y, \\ \dot{y} = x + 2e^t. \end{cases}$

782. $\begin{cases} \dot{x} = 4x - 3y + \sin t, \\ \dot{y} = 2x - y - 2\cos t. \end{cases}$ **783.** $\begin{cases} \dot{x} = 2x + y + 2e^t, \\ \dot{y} = x + 2y - 3e^{1t}. \end{cases}$

784. $\begin{cases} \dot{x} = x - y + 8t, \\ \dot{y} = 5x - y. \end{cases}$ **785.** $\begin{cases} \dot{x} = 2x - y, \\ \dot{y} = 2y - x - 5e^t \sin t. \end{cases}$

In problems 786-790 solve the given systems by the method of variations of constants.

786. $\begin{cases} \dot{x} = y + \operatorname{tg}^2 t - 1, \\ \dot{y} = -x + \operatorname{tg} t. \end{cases}$ **787.** $\begin{cases} \dot{x} = 2y - x, \\ \dot{y} = 4y - 3x + \dfrac{e^{3t}}{e^{2t}+1}. \end{cases}$

788. $\begin{cases} \dot{x} = -4x - 2y + \dfrac{2}{e^t - 1}, \\ \dot{y} = 6x + 3y - \dfrac{3}{e^t - 1}. \end{cases}$

789. $\begin{cases} \dot{x} = x - y + \dfrac{1}{\cos t}, \\ \dot{y} = 2x - y. \end{cases}$ **790.** $\begin{cases} \dot{x} = 3x - 2y, \\ \dot{y} = 2x - y + 15e^t \sqrt{t}. \end{cases}$

Section 15

LYAPUNOV STABILITY

Foundations of Lyapunov's stability theory are to be found in several books. See for example, the Introduction of: Krasovskii, The Theory of Motion, Stanford University Press, 1963; Coddington and Levinson, Differential Equations, McGraw-Hill, 1952; Lefschetz and LaSalle, Stability Theory of Differential Equations, Prentice-Hall, 1962.

791. Use the definition of Lyapunov's stability to decide whether the solution of the equation $dx/dt = t - x$, with initial value $x(0) = 1$, is stable or unstable.

792. Answer the same question for the solution of the system

$$dx/dt = 4y, \quad dy/dt = -x$$

with initial conditions $x(0) = 0$, $y(0) = 0$.

In problems 793-796 use Lyapunov's theorem on first order stability to determine whether the trivial solution $x(t) = 0$, $y(t) = 0$ is a stable solution of the given system.

793. $\begin{cases} \dot{x} = 2xy - x + y, \\ \dot{y} = 5x^4 + y^3 + 2x - 3y. \end{cases}$ **794** $\begin{cases} \dot{x} = x^2 + y^2 - 2x, \\ \dot{y} = 3x^2 - x + 3y. \end{cases}$

795. $\begin{cases} \dot{x} = e^{x+2y} - \cos 3x, \\ \dot{y} = \sqrt{4 + 8x} - 2e^y. \end{cases}$ **796.** $\begin{cases} \dot{x} = \ln(4y + e^{-3x}), \\ \dot{y} = 2y - 1 + \sqrt[3]{1 - 6x}. \end{cases}$

In problems 797-800 find the equilibrium point and determine whether it is a stable or unstable solution.

797. $\begin{cases} \dot{x} = y - x^2 - x, \\ \dot{y} = 3x - x^2 - y. \end{cases}$ **798.** $\begin{cases} \dot{x} = (x - 1)(y - 1), \\ \dot{y} = xy - 2. \end{cases}$

799. $\begin{cases} \dot{x} = 5 - x^2 - y^2, \\ \dot{y} = 1 + y^2 - x. \end{cases}$ **800.** $\begin{cases} \dot{x} = y, \\ \dot{y} = \sin(x + y). \end{cases}$

801. Determine whether the solution $x = t$, $y = -t^2$ is
a stable solution of the system.

802. A certain system of equations has for solutions the
family

$$ x = \frac{C_1 - C_2 t}{1 + t^2}, \qquad y = (C_1 t^3 + C_2)e^{-t}. $$

Determine whether the null-solution $x(t) = 0$, $y(t) = 0$ is a
stable solution of the system.

803. Suppose the functions $P(x,y)$, $Q(x,y)$ are continuous
and have continuous first derivatives, and furthermore reduce
to zero at the origin. Suppose further that the trajectories
of the system $dx/dt = P(x,y)$, $dy/dt = Q(x,y)$ have the form

given in equation (1). What can be said about the behavior of every solution for $t \to +\infty$? Is the origin an asymptotically stable solution? Is there any stable solution of the system?

804. Suppose $a(t)$ is a continuous function of t. Show that a necessary and sufficient condition that the null solution of the equation

$$dx/dt = a(t) \, x$$

be stable is that the relation

$$\overline{\lim_{t \to +\infty}} \int_0^t a(s)ds < +\infty.$$

hold.

805.* Show that if any particular solution of a linear system of differential equations is stable, then every solution of the same system is stable.

In problems 806-810, use a Lyapunov function to settle the stability of the null solution of the given system. See the Krasovskii reference cited at the beginning of this section.

806. $\begin{cases} \dot{x} = x^3 - y, \\ \dot{y} = x + y^3. \end{cases}$ 807. $\begin{cases} \dot{x} = y - x + xy, \\ \dot{y} = x - y - x^2 - y^3. \end{cases}$

808. $\begin{cases} \dot{x} = 2y^3 - x^5, \\ \dot{y} = -x - y^3 + y^5. \end{cases}$ 809. $\begin{cases} \dot{x} = y - 3x - x^3, \\ \dot{y} = 6x - 2y. \end{cases}$

810*. $\begin{cases} \dot{x} = -f_1(x) - f_2(y), \\ \dot{y} = f_3(x) - f_4(y). \end{cases}$

where sgn $f_i(z) = $ sgn z , $i = 1, 2, 3, 4.$

Section 16

SINGULAR POINTS

If the functions $P(x,y)$, $Q(x,y)$ have continuous derivatives with respect to both arguments, the points (x,y) that
verify both the equations $P(x,y) = 0$, $Q(x,y) = 0$ are called
<u>singular</u> <u>points</u> of the equation

$$dy/dx = P(x,y)/Q(x,y) \qquad (1)$$

or of the system

$$dx/dt = P(x,y), \quad dy/dt = Q(x,y) . \qquad (2)$$

Consider the system

$$dx/dt = ax + by , \quad dy/dt = cx + dy, \qquad (3)$$

or the single equation

$$dy/dx = (cx + dy)/(ax + by) , \qquad (4)$$

where a,b,c,d are real constants.

First suppose that the determinant

$$\det \begin{bmatrix} a & b \\ c & d \end{bmatrix} \qquad (5)$$

does not have the value zero.

Then the singular points can be characterized by examining the characteristic roots, that is the solutions of the determinantal equation

$$\det \begin{bmatrix} a - \lambda & , & c \\ & & \\ b & , & d - \lambda \end{bmatrix} = 0 .$$

    i. If the characteristic roots are real, distinct, and of the same sign, the singular point is a node (fig. 2i).

    ii. If the characteristic roots are real and distinct but of opposite sign, the singular point is a saddle point (fig. 2ii).

    iii. If the characteristic roots are neigher real nor pure imaginary, the singular point is a focus (fig. 2iii).

    iv. If the characteristic roots are pure imaginary, the singular point is a center (fig 2iv).

    v. If the characteristic roots are equal, but nonzero, the singular point can be a degenerate node or a singular node. The latter case occurs if the equations have the form $dx/dt = ax$, $dy/dt = ay$ (or $dy/dx = y/x$); see fig. 2vi. The former case occurs otherwise. See fig. 2v.

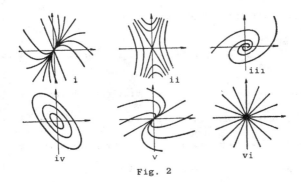

Fig. 2

The case when the determinant ad - bc is zero does not really require discussion. In this case cx + dy is proportional to ax + by, and the equation has the form dy/dx = k, or dx/dy = k', and the trajectories are parallel lines.

The family of trajectories in the plane is conveniently drawn by marking first those trajectories that are straight lines and pass through the node, saddle, or degenerate node. For a focus, it is necessary to find the direction the trajectories take in winding around the singular point. These ideas are illustrated in the examples below.

Example 1. Find the nature of the singular point of the equation

$$dy/dx = (x + y)/(2x). \qquad (6)$$

The characteristic equation is

$$\det \begin{bmatrix} 2 - \lambda\,, & 0 \\ 1 & ,\ 1 - \lambda \end{bmatrix} = 0; \quad (2 - \lambda)(1 - \lambda) = 0; \quad \lambda = 1, \quad \lambda = 2.$$

The characteristic roots are real, distinct, and of like sign. Thus the singular point is a node, fig. 2i. The lines that are trajectories passing through the singular point have the form $y = kx$, and if we substitute this in (6), we obtain

$$k = (x + kx)/(2x), \quad 2k = 1 + k, \quad k = 1.$$

This gives the line $y = x$, but there should be two lines, since the node is not degenerate. The missing line must be the only one that goes through (0,0) and does not have an equation of the form $y = kx$, namely the line $x = 0$. This latter is indeed a solution of the equation $dx/dy = 2x/(x + y)$, or of the system $dx/dt = 2x$, $dy/dt = x + y$.

The entire system of trajectories can be sketched by drawing the graph of the differential equation. For example, at every point on the line $y = -3x$, y' has the value -1, and the line elements drawn on this line show the direction taken by trajectories that intersect it. Similarly on the line $y = -3x$ and $x = 0$ cannot leave this sector. Therefore they are

all tangent to the line x = 0, and the trajectory curves are
as in fig. 3b.

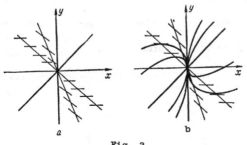

Fig. 3

Example 2. Find the character of the singular point of
the equation

$$\frac{dy}{dx} = \frac{4x - 3y}{x - 2y} \ . \tag{7}$$

The characteristic equation in this case is

$$\det \begin{bmatrix} 1 - \lambda, & -2 \\ 4, & -3 - \lambda \end{bmatrix} = 0, \ \lambda^2 + 2\lambda + 5 = 0; \ \lambda = -1 \pm 2 \ i.$$

The singular point is a focus. It is desirable to consider
rather than (7), the system

$$dx/dt = x - 2y, \quad dy/dt = 4x - 3y.$$

Let us draw the velocity vector (dx/dt, dy/dt) in the point
(1,0). By (8), the formula for its components is

90

$(x - 2y, 4x - 3y)$, that is $(1,4)$. See fig. 4a.

Thus as t increases, the direction of motion is as indicated
by the arrow. Since the real part of the characteristic roots
is -1 (i.e. is negative), the focus is asymptotically stable;
thus the trajectories wind tighter and tighter about the singular point as t increases. The integral curves have the form
of fig. 4b.

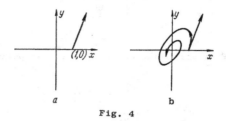

Fig. 4

For equations (1) or (2), the method of studying singular
points is to expand the functions $P(x,y)$, $Q(x,y)$ in a Taylor
series about the singular point. System (1) then takes the form

$$\frac{dx_1}{dt} = ax_1 + by_1 + \phi(x_1, y_1), \quad \frac{dy_1}{dt} = cx_1 + dy_1 + \psi(x_1, y_1). \quad (9)$$

where x_1, y_1 are the new (translated) coordinates $x - h$, $y - k$,
and a,b,c,d are constants. The case we can study is that in
which the relations

$$\frac{\phi(x_1, y_1)}{r^{1+\epsilon}} \to 0, \quad \frac{\psi(x_1, y_1)}{r^{1+\epsilon}} \to 0$$

hold for $x \to 0$, $y \to 0$, where $r = \sqrt{x^2 + y^2}$.

This condition is clearly satisfied for $0 < \epsilon < 1$, if the functions P,Q are twice differentiable at the point (h,k). If all the characteristic roots of equation (5) have nonzero real part, the singular point of system (9) will have the same type as the singular point of the corresponding system (3) that is obtained by neglecting the functions ϕ, ψ in (9). Indeed the trajectories of equation ·(9) approach those of equation (3) as a point moves along a trajectory towards the singular point. However, to the line $y = kx$ in (3) there may well correspond a slightly distorted line (a curve) from equation (9). The curves of the one system also approach those of the other when the singular point is a focus.

However, when the singular point of (3) is a center, the singular point of (9) can be either a center or a focus (usually the latter). If the singular point of (9) is asymptotically stable for $t \to \infty$ or $t \to -\infty$, it will be a focus. If a Lyapunov function can be constructed to settle

this question, the singular point will be completely described.

If the integral curves of (9) have an axis of symmetry passing

through the singular point, and if that point is a center for

(3), then it will be a center for (9). (This condition is

sufficient, but not necessary.) In particular, there is an

axis of symmetry if P(x,y), Q(x,y) are unchanged when x is

replaced by 2h - x, or if they are unchanged when y is re-

placed by 2k - y.

Find the character of the singular points in problems

811-828. Sketch the integral curves in the (x,y)-plane.

811. $y' = \dfrac{2x+y}{3x+4y}$.   **812.** $y' = \dfrac{x-4y}{2y-3x}$.

813. $y' = \dfrac{y-2x}{y}$.   **814.** $y' = \dfrac{x+4y}{2x+3y}$.

815. $y' = \dfrac{x-2y}{3x-4y}$.   **816.** $y' = \dfrac{2x-y}{x-y}$.

817. $y' = \dfrac{y-2x}{2y-3x}$.   **818.** $y' = \dfrac{4y-2x}{x+y}$.

819. $y' = \dfrac{y}{x}$.   **820.** $y' = \dfrac{4x-y}{3x-2y}$.

821. $\begin{cases} \dot{x} = 3x, \\ \dot{y} = 2x + y. \end{cases}$ **822.** $\begin{cases} \dot{x} = x + 2y, \\ \dot{y} = 5y - 2x. \end{cases}$

823. $\begin{cases} \dot{x} = x + 3y, \\ \dot{y} = -6x - 5y. \end{cases}$ **824.** $\begin{cases} \dot{x} = x, \\ \dot{y} = 2x - y. \end{cases}$

825. $\begin{cases} \dot{x} = -2x - 5y, \\ \dot{y} = 2x + 2y. \end{cases}$ **826.** $\begin{cases} \dot{x} = 3x + y, \\ \dot{y} = y - x. \end{cases}$

827. $\begin{cases} \dot{x} = 3x - 2y, \\ \dot{y} = 4y - 6x. \end{cases}$ **828.** $\begin{cases} \dot{x} = y - 2x, \\ \dot{y} = 2y - 4x. \end{cases}$

Find both the singular points and their character in problems 829-832.

829. $y' = \dfrac{2y - x}{3x + 6}$.

830. $y' = \dfrac{2x + y}{x - 2y - 5}$.

831. $\begin{cases} \dot{x} = 2x - y, \\ \dot{y} = x - 3 \end{cases}$

832. $\begin{cases} \dot{x} = x + y - 1, \\ \dot{y} = x - y - 3. \end{cases}$

Sketch the integral curves of systems 833-842 in the neighborhood of each singular point. First, state the coordinates of the singular point, and its nature.

833. $y' = \dfrac{6x - y^2 + 1}{2x + y^2 - 1}$.

834. $y' = \dfrac{2y - 2}{4y^2 - x^2}$.

835. $y' = \dfrac{4y^2 - x^2}{2xy - 4y - 8}$.

836. $y' = \dfrac{xy}{4 - 4x - 2y}$.

837. $y' = \dfrac{2y}{x^2 - y^2 - 1}$.

838. $y' = \dfrac{2x}{1 - x^2 - y^2}$.

839. $y' = \dfrac{2x(x - y)}{2 + y - x^2}$.

840. $y' = \dfrac{2xy}{1 - x^2 - y^2}$.

841. $y' = \dfrac{y^2 - x^2}{2(x - 1)(y - 2)}$.

842. $y' = \dfrac{x(2y - x + 5)}{x^2 + y^2 - 6x - 8y}$.

Sketch the integral curves for problems 843-847 in the neighborhood of the origin. Note that in these cases, the singular point is not of the type considered in the discussion at the beginning of this section. Therefore it will be necessary to draw some isoclines. It will also be necessary to determine the limiting directions of the integral curves at the singular points.

843*. $y' = \dfrac{xy}{x+y}$.   844*. $y' = \dfrac{x^2+y^2}{x^2+y}$.

845*. $y' = \dfrac{xy}{y+x^2}$.   846*. $y' = \dfrac{xy}{y-x^2}$.

847*. $y' = \dfrac{y^2}{y+x^2}$.

848. If the singular point of the curve

$$(ax + by)\, dx + (mx + ny)\, dy = 0$$

is a center, show that the equation is an exact differential.
The converse is invalid.

849.* If the equation of the preceding exercise is not
exact and has no integrating factor in the neighborhood of
the origin, then the singular point is a saddle point.

850.* Suppose that the functions $p(x,y)$, $q(x,y)$ are
defined and continuously differentiable in some neighborhood
of the origin $(0,0)$, and that each of these functions and each
of its first derivatives is 0 at the origin. Show that, if
the equation

$$y' = \frac{ax + by + p(x,\, y)}{cx + dy + q(x,\, y)} \tag{10}$$

is unaltered by the substitution $y \to -y$, and if the roots of
the characteristic equation

$$\det \begin{bmatrix} c - \lambda, & d \\ a & , b - \lambda \end{bmatrix} = 0$$

are pure imaginaries, then the point $(0,0)$ is a center.

Section 17

PROBLEMS IN THE THEORY OF OSCILLATIONS

Practically all problems in this section are derived from
or based on physical phenomena, and lead to linear equations or
systems with constant coefficients. In case the problems lead
to equations that cannot be solved by use of methods in the
preceding sections, the trajectories should be sketched on the
phase plane, and the particular solution should be tested for
stability.

Problems 862-868, 873 are from the theory of electric cir-
cuits. The following facts are needed. At every point, in particular at every node or junction of a circuit, the sum of the
negative (inflowing) currents is equal to the sum of the positive
(outflowing) currents. A voltage jump occurs at a device, like
a battery or generator, for producing voltage. In every closed
loop of a network, the algebraic sum of the voltage jumps is
equal to the algebraic sum of the voltage drops. These voltage
drops are computed as follows.

The voltage drop across a resistor of R ohms is RI volts, where I is the current in amperes. The voltage drop across an inductor of inductance L henrys is L·dI/dt, where t is the time in seconds. The voltage drop across a capacitor of capacity C farads is q/C, where q = q(t) is the charge in coulombs on the capacitor at time t . Moreover, I = dq/dt, and I = I(t) in the above examples is the current strength flowing through the circuit in question at time t.

Example. The voltage E = V sin wt is applied to a circuit consisting of a resistor of resistance R ohms and a capacitor of capacity C farads in series. Find, as a function of time, the "current neglecting transients," i.e., find the current, assuming that the current is a constant or a periodic function of the time.

Solution. The explanation above shows that at time t, the current I(t) at each point of the circuit is the same as that at every other point. The voltage drops are RI and q/C through the resistor and capacitor respectively. Therefore the voltage relation

$$RI + q/C = V \sin wt$$

holds. We differentiate this relation with respect to t, and replace dq/dt by I, thus obtaining

$$R \cdot dI/dt + I/C = V w \cos wt . \qquad (1)$$

This is the differential equation to solve. It is a linear equation with constant coefficients; examples of this type appear in section 11. It is clear that this equation will have a periodic solution, of the form

$$I(t) = A_1 \cos wt + B_1 \sin wt . \qquad (2)$$

In fact, if the latter expression is substituted into (1), the constants A_1 and B_1 are easily determined.

However, in electrical engineering, it is customary to write a periodic current in a form that makes its maximum value apparent. Thus we write (2) in the form

$$I = A \sin (wt - \emptyset) \qquad (3)$$

which is actually equivalent to (2). After we substitute (3) into (1), we expand the trigonometric functions of wt - ∅ so that only functions of wt and of ∅ appear. Next we collect the terms in sin wt, cos wt and compare coefficients. This yields two equations, as follows

$$RAw \sin \emptyset + (A/C) \cos \emptyset = 0,$$

$$RAw \cos \emptyset - (A/C) \sin \emptyset = Vw$$

Thus we obtain

$$\tan \emptyset = -1/(RCw), \quad A = \frac{V}{\sqrt{R^2 + \left(\frac{1}{wC}\right)^2}}$$

We have obtained the "current neglecting transients;" this current $I(t)$ is sometimes called the steady-state current. A complete solution of equation (1) is obtained by adding to the function $I(t)$ already obtained the general solution of the homogeneous (reduced) equation

$$R \cdot dI/dt + I/C = 0 . \qquad (4)$$

The general solution of (4) is $I = K \exp(-t/RC)$, where K is an arbitrary constant. But for $t \to \infty$ this part of the solution approaches 0 exponentially, and thus every solution of equation (1) approaches the steady state solution; this explains the use of the term.

The above equations for $\emptyset$, A clearly have a geometric interpretation, which is often used in electrical engineering.

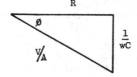

It is important to be able to draw a locus showing the
variations of q with I, or of I with dI/dt, the time
being a parameter. This locus will be a curve in the (q,I)-
plane or in the (I,İ)- plane, See the books by Coddington-
Levinson, Lefschetz-LaSalle, Andronow-Chaikin.

Limit cycles. Suppose a trajectory is a closed curve; it
therefore represents periodic (or possibly damped deadbeat)
motion. If some neighborhood of this closed trajectory consists
entirely of trajectories that approach the given trajectory as
a limit for $t \to \infty$, the closed trajectory is called a stable
limit cycle. If the approach is similar, but for $t \to -\infty$, the
closed trajectory is called an unstable limit cycle. If on one
side of the trajectory, the trajectories in the neighborhood
approach the given trajectory for $t \to \infty$, and on the other side
of the given trajectory, the trajectories in the neighborhood
approach the given trajectory for $t \to -\infty$, the trajectory
is called semistable.

851. A weight of mass m is attached to one end of a
spring, and the other end is fixed. The spring constant is
k, i.e. if the weight is displaced from equilibrium a distance

x, the spring exerts a force of kx gm. towards equilibrium. Moreover, if the weight moves with speed v, there is a resistance of nv grams, where n is a constant. At time t = 0, the weight is placed in the equilibrium position and propelled with velocity v_o in the direction in which the spring acts. Find the motion of the weight in the two cases $n^2 < 4 k m$, $n^2 > 4 k m$.

852.* Suppose that m and k are given in the preceding problem. Determine n so that the weight reaches its equilibrium position in the shortest possible time (that is, so that the solution x(t) approaches its equilibrium value as quickly as possible, t → ∞.

Draw the trajectories for problems 853-859 in the phase plane. Explain the behavior of the solutions for t → ∞, by reference to the sketch, or on the sketch.

853. $\ddot{x} + 4x = 0.$ **854.** $\ddot{x} - 4\dot{x} + 3x = 0.$
855. $\ddot{x} + 2\dot{x} + 5x = 0.$ **856.** $\ddot{x} - \dot{x} - 2x = 0.$
857. $2\ddot{x} + 5\dot{x} + 2x = 0.$ **858.** $\ddot{x} + 2\dot{x} + x = 0.$
859. $\ddot{x} - 2\dot{x} + 2x = 0.$

860. Two pulleys are attached to a shaft; they have moments of inertia I_1, I_2. To twist one of the pulleys an angle ϕ with respect to the other requires an elastic shaft-deforming torque of $K\phi$. Find the frequency of the torsional oscillations of the shaft.

861. A weight of mass m is attached to one end of an elastic rod. The other end moves so that its position at time t has coordinate $B \sin wt$. The elastic force exerted by the rod is proportional to the difference in displacements of its ends. Neglecting the mass of the rod, and friction, find the amplitude A of the forced vibrations of the mass. Can the relation $A > B$ hold?

862. An electric circuit consists of a voltage source that supplies voltage V volts, a resistor of resistance R ohms, and an inductance of L henrys, together with a switch that is closed at time $t = 0$. Find the current as a function of the time.

863. Solve the preceding problem, replacing the inductance L by a capacitor of capacity C farads. The capacitor is uncharged when the switch is closed.

864. A resistor of resistance R ohms is connected to a capacitor of capacity C farads that has a charge q coulombs at time t = 0. The circuit is closed at t = 0. Find the current as a function of time for t > 0.

865. An inductor, resistor, and capacitor are connected in series. At time t = 0, the circuit is closed, the capacitor having a charge of q coulombs at that time. Find the current as a function of time, and the frequency of current change in case the current does actually change sign periodically.

866. A voltage source supplies voltage E = V sin wt in a circuit consisting of the voltage source, and a resistor and capacitor in series. Find the steady-state current in the circuit.

867. A voltage source supplies voltage E = V sin wt and is connected to a resistor, inductor, and capacitor in series. Find the steady-state current in the circuit. What frequency w is needed to obtain the maximum possible current?

868. See figure

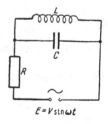

$E = V \sin \omega t$

A voltage source supplies voltage $E = V \sin wt$, through a resistor of resistance R ohms. The circuit then splits into two branches consisting of an inductor and capacitor (in parallel), as shown. Find the steady-state current through the resistor. What frequencies result in non-oscillatory currents? What frequency results in the least current?

869.[*] Let the function $f(t)$ have period T, and maximum value $\max |f(t)| = m$. Find (in the form of a definite integral) the periodic solution(s) of the equation

$$\ddot{x} + 2 b \dot{x} + x = f(t),$$

and give an upper bound to the amplitude of the periodic solution(s).

870. A weight of mass $m = 2$ is attached to a spring like that of problem 851, with spring constant $k = 2$. At time $t = 0$, the weight is released from position $h = 5$ above the equilibrium position, with velocity 0. What is the greatest excursion of the weight to the other side of the equilibrium position, if the frictional force $f = 1$ is independent of the velocity? Describe the subsequent motion of the weight, and draw the trajectory in the phase plane.

871. Write the differential equation of motion for a pendulum not acted on by frictional or dissipative forces. Take the case that the values of the coefficients in the equation are all 1, and draw the trajectories in the phase plane. Explain the physical significance of the various types of trajectory.

872. Find the differential equation of motion for a pendulum swinging under resistance proportional to the square of the velocity. Sketch the trajectories on the phase plane.

Hint. Use the sketch from problem 871.

873.

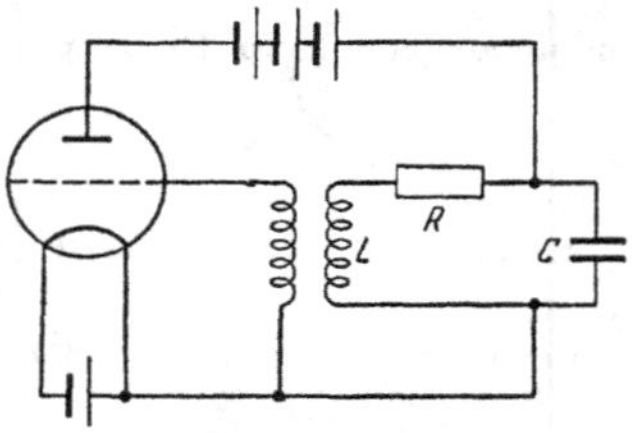

Fig. 6

Give the equation for the amount of current I_L that passes through the capacitor L in a simple oscillating vacuum tube circuit. In the tube, the upper element is the plate or cathode, the central element is the (perforated) grid, and the lower element is the filament or anode. Assume that the plate current I_a (from anode to cathode) is a known function

$I_a = f(v)$ of the voltage V on the grid, and that the function

F is given graphically, as in this figure.

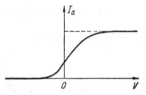

Suppose further that the mutual inductance between the two in-

ductances is M, and that the voltage V is given (to within

sign) by the relation

$$V = \pm M \frac{d\,I_L}{dt}.$$

The oscillating circuit in figure 6 contains an inductance, re-

sistance, and capacitance. Under what conditions is the equili-

brium condition $I_L(t)$ = const. an unstable condition?

874. Indicate the isoclines, and sketch the trajectories

of the equation in problem 873 in the phase plane, in case the

equilibrium I_L = const. is unstable. Find the behavior of

the solutions for $t \to +\infty$.

Hint. Put the equation into the form

$$\ddot{x} + F(\dot{x}) + x = 0,$$

and sketch the graph of the function F.

In problems 875-884, find the singular points by drawing
isoclines and sketching the trajectories in the phase plane.
From the sketch, describe the behavior of the solutions for
$t \to +\infty$, and state whether periodic solutions of the equation
can exist.

The second term in problem 884 is $2^{\overset{*}{x}}$.

875. $\ddot{x} + 2\dot{x} + \dot{x}^2 + x = 0.$ **876.** $\ddot{x} - 5\dot{x} - 4x + x^2 = 0.$
877. $2\ddot{x} - \dot{x}^2 - x^2 - 2x = 0.$ **878.** $\ddot{x} + \dot{x} + 2x - x^2 = 0.$
879. $\ddot{x} + \dot{x}^2 - x^2 + 1 = 0.$ **880.** $2\ddot{x} + \sin x - \sin 2x = 0.$
881. $\ddot{x} + \dot{x}^3 - \dot{x} + x = 0.$ **882.** $\ddot{x} + (x^2 - 1)\dot{x} + x = 0.$
883. $\ddot{x} + \dot{x} - 2 \operatorname{arctg} \dot{x} + x = 0.$
884. $\ddot{x} + 2^{\dot{x}} - \dot{x} + x = 0.$

Without carrying out the integration to explicit formulas,
describe the solutions of equations 885-891, which are given in
polar coordinates (r, ϕ), and determine whether limit cycles
exist.

885. $\dfrac{dr}{d\varphi} = r(1 - r^2).$ **886.** $\dfrac{dr}{d\varphi} = r(r-1)(r-2).$
887. $\dfrac{dr}{d\varphi} = r(1-r)^2.$ **888.** $\dfrac{dr}{d\varphi} = \sin r.$
889. $\dfrac{dr}{d\varphi} = (|r-1| - |r-2| - 2r + 3)\, r.$
890. $\dfrac{dr}{d\varphi} = r \sin \dfrac{1}{r}.$ **891.** $\dfrac{dr}{d\varphi} = r(1-r)\sin \dfrac{1}{1-r}.$

892.* Determine the values of the constant a for which
the system

$$\frac{d\varphi}{dt} = 1, \quad \frac{dr}{dt} = (r-1)(a + \sin^2 \varphi)$$

has stable limit cycles, and the values of a for which it has

unstable limit cycles.

893.* Let f(r) be a continuous function. Determine
conditions under which the system of equations

$$\frac{dr}{dt} = f(r), \quad \frac{d\phi}{dt} = 1,$$

has limit cycles. When will these limit cycles be stable, un-

stable, semistable?

894.* Let F(z) be a continuous function of z with the
properties that, for z > 0 the relation F(z) > F(0) holds,

and for z < 0 the relation F(z) < F(0) holds. Show that the

equation

$$\ddot{x} + F(\dot{x}) + x = 0$$

cannot have a limit cycle (in the phase plane).

DEPENDENCE OF SOLUTIONS ON INITIAL CONDITIONS
AND ON PARAMETERS

APPROXIMATE SOLUTION OF DIFFERENTIAL EQUATIONS

Let $x_1(t), \ldots, x_n(t)$ be the solution of the system

$$dx_i/dt = f_i(t, x_1, \ldots, x_n), \quad i = 1, \ldots, n, \tag{1}$$

where all the functions f_i are differentiable and the relations $|\partial f_i / \partial x_j| \le K$ hold for $i, j = 1, \ldots, n$. Suppose that the functions $y_1(t), \ldots, y_n(t)$ are functions that satisfy the inequalities

$$\left| dy_i/dt - f_i(t, y_1, \ldots, y_n) \right| \le \eta_i, \quad i = 1, \ldots, n, \tag{2}$$

$$\left| y_i(0) - x_i(0) \right| \le \delta_i, \quad i = 1, \ldots, n. \tag{3}$$

Then the estimate below holds.

$$\sum_{i=1}^{n} \left| x_i(t) - y_i(t) \right| \le \delta \, e^{Kn|t|} + [\eta/Kn] \, (e^{Kn|t|} - 1)$$

where $\delta = \delta_1 + \cdots + \delta_n$, $\eta = \eta_1 + \ldots + \eta_n$.

This latter inequality can be used to estimate the degree with which the functions $y_1, \ldots, y_n$ approximate a solution of the system (1), and also the difference between a solution of the system (1) and the system

$$dy_i/dt = g_i(t, y_1, \ldots, y_n), \quad i = 1, \ldots, n,$$

if it is known that the relations $\left| g_i - f_i \right| \leq \eta_i$ hold for

$i = 1, \ldots, n$.

We now consider the system

$$dx_i/dt = f_i(t, x_1, \ldots, x_n, \mu), \quad i = 1, \ldots, n \quad (4)$$

which involves a parameter μ and which is to be solved subject

to the initial conditions

$$x_i(0) = a_i(\mu), \quad i = 1, \ldots, n \quad (5)$$

where the parameter μ usually takes on values which do not

vary greatly from a neutral value, and the functions f_i, a_i

are continuous for $i = 1, \ldots, n$, and have continuous derivatives

with respect to $x_1, \ldots, x_n$, μ. The hypotheses just stated

insure that the solution of (4) will be a continuously differ-

entiable function of the parameter μ. The derivative

$\partial x_i / \partial \mu = u_i$, satisfies the linear system of equations

$$du_i/dt = \sum_{j=1}^{n} (\partial f_i / \partial x_j) u_j + \partial f_i / \partial \mu, \quad i = 1, \ldots, n, \quad (6)$$

subject to the initial conditions

$$u_i(0) = a_i'(\mu), \quad i = 1, \ldots, n.$$

In formula (6) the values of the derivatives $\partial f_i / \partial x_j$, $\partial f_i / \partial \mu$

are to be computed for $x_1 = x_1(t), \ldots, x_n(t)$, where the functions represent a solution of system (4) subject to the initial conditions (5).

In particular if the initial conditions are taken in the form $a_k(\mu) = \mu$, $a_i(\mu)$ = const for values of $i \neq k$, and if we assume that all the functions $f_1, \ldots, f_n$ are independent of μ, the assertion above can be particularized. It states in this case that the system (4), subject to the initial conditions $x_i(0) = a_i(\mu)$, $i = 1, \ldots, k, \ldots, n$ will have a solution that varies with the initial condition as follows. The derivative $\partial x_i / \partial a_k = u_i$ of the arbitrary component x_i with respect to the initial value a_k is obtainable from the formula

$$du_i/dt = \sum_{j=1}^{n} \partial f_i / \partial x_j, \quad i = 1, \ldots, n.$$

The initial conditions for the latter equation are $u_i(0) = 0$ for $i \neq k$, $u_k(0) = 1$.

895. The equation $y' = x + \sin y$ is to be solved on the interval (0, 1) subject to the initial condition $y(0) = y_o = 0$. If y_o is perturbed by 0.01, estimate the maximum variation of the solution on this interval.

896. The equation $\ddot{x} + \sin x = 0$ is to be solved on the interval $0 \leq t \leq T$ subject to the initial conditions

$x(0) = 0$, $\dot{x}(0) = 0$. If the right member is replaced by a function $\phi(t)$ that has absolute value ≤ 0.1, $|\phi(t)| \leq 0.1$, how much will the solution be perturbed? The question concerns the action of a small perturbing force.

897. The system $\dot{x} = x-y$, $\dot{y} = tx$ is to be solved for the initial conditions $x(0) = 1$, $y(0) = 0$, on the interval $|t| \leq 0.1$. Find the maximum error in the solution

$$\tilde{x}(t) = 1 + t + \frac{1}{2}t^2, \quad \tilde{y}(t) = \frac{1}{2}t^2.$$

898. The equation $y'' - x^2y = 0$ is to be solved on the interval $|x| \leq 0.5$, subject to the initial conditions $y(0) = 1$, $y'(0) = 0$. Find the error in using the approximate solution

$$\tilde{y}(x) = \exp(x^4/12).$$

899.[*] Answer the same question for the differential equation $y' = 2x\,y^2 + 1$ on the interval $|x| \leq 0.25$ subject to the initial condition $y(0) = 1$ for the solution

$$\tilde{y}(x) = (1-x)^{-1}.$$

In problems 900-902 find the indicated derivative with respect to the parameter or with respect to the initial condition.

900. $y' = y + \mu(x + y^2)$, $y(0) = 1$; find $\dfrac{\partial y}{\partial \mu}\Big|_{\mu=0}$.

901. $\begin{cases} \dot{x} = xy + t^2 \\ 2\dot{y} = -y^2 \end{cases}$ $x(1) = x_0 = 3$, $y(1) = y_0 = 2$; find $\dfrac{\partial x}{\partial y_0}$.

902. $\ddot{x} - \dot{x} = (x+1)^2 - \mu x^2$; $x(0) = \dfrac{1}{2}$, $\dot{x}(0) = -1$;

find $\dfrac{\partial x}{\partial \mu}\Big|_{\mu=1}$.

903. * Suppose the differential equation $\ddot{x} = f(t, x, \dot{x})$ has a solution on the interval $0 \le t \le T$ subject to the initial conditions $x(0) = a$, $\dot{x}(0) = b$. Suppose further that the function $f(t, x, y)$ has continuous derivatives, and that $\partial f / \partial x \ge 0$. Demonstrate that on the entire t - interval, the derivative of the solution with respect to the initial value b is positive.

904. * Suppose the system of differential equations

$$dx_i/dt = f_i(t, x_1, \ldots x_n), \quad i = 1, \ldots, n$$

has a solution for $t \ge t_o$ subject to the initial conditions $x_i(t_o) = x_{io}$. Suppose further that the functions f_i and the partial derivatives $\partial f_i / \partial x_k$ are continuous and that the inequalities $|\partial f_o / \partial x_k| \le L(t)$, hold for $i, k = 1, \ldots, n$.

Establish the inequalities

$$|\partial x_i / \partial x_{k0}| \le \exp n \int_{t_o}^{t} L(s)\, ds, \quad i, k = 1, \ldots, n.$$

Use the method of small parameters in problems 905-914 to find a periodic function which is an approximate solution of the given equations. If the right member is a periodic function

of t, the solution must have the same period as the right mem-

ber. The symbol μ is a small parameter. The method desired is

explained in the book of Coddington and Levinson.

905. $\ddot{x} + 3x = 2\sin t + \mu \dot{x}^2$.   **906.** $\ddot{x} + 5x = \cos 2t + \mu x^2$.
907. $\ddot{x} + 3x + x^3 = 2\mu \cos t$.   **908.** $\ddot{x} + x^2 = 1 + \mu \sin t$.
909. $\ddot{x} + \sin x = \mu \sin 2t$.

910. $\ddot{x} + x - \sin 3t - \sin 2t + \mu \ x^2$;   only find the zeroth
approximation.

911*. $\ddot{x} + x = 6\mu \sin t - x^3$.   **912.** $\ddot{x} + x - x^2 = 0$.
913. $\ddot{x} + \sin x = 0$.   **914.** $\ddot{x} + x = \mu(1 - x^2)\dot{x}$.

Use Adam's or Störmer's method in problems 915-917 to ob-

tain an approximate solution of the equations on the interval

indicated. Carry three decimal places (accuracy of 0.001). Use

a power series to obtain starting values. Descriptions of the

methods of approximate calculation are to be found in books on

numerical analysis, e.g. Hildebrand, F.B. Introduction to numer-

ical analysis, New York: McGraw-Hill, 1956, p. 198; Collatz, L.

Numerische Behandlung von Differentialgleichungen, Berlin:

Springer, 1955, pp. 79, 81, 499.

915. $y' = y,$ $y(0) = 1;$ $0 \leqslant x \leqslant 1.$
916. $y' = y^2 - x,$ $y(0) = 0.5;$ $0 \leqslant x \leqslant 1.$
917. $xy'' + y' + xy = 0,$ $y(0) = 1,$ $y'(0) = 0;$ $0 \leqslant x \leqslant 1.$

Problems 918-920 can be solved by equating the slope of the direction field attached to the equation $y' = f(x,y)$ with the slope of appropriately chosen curves $y = \phi_i(x)$.

918. Show that the solution of the equation $y' = x - y^2$ that satisfies the initial condition $y(4) = 2$ satisfies the inequalities $\sqrt{x} - 0.07 < y(x) < \sqrt{x}$ for all positive values of x exceeding 4, $4 < x$.

919.* Show that the solution $y(x)$ of the equation $y' = x - y^2$ that satisfies the initial condition $y(x_0) = y_0'$, $x_0 \geq 0,$ $y_0 \geq 0$ satisfies the relation $y(x) - \sqrt{x} \to 0$ for $x \to \infty$.

920.* Give upper and lower estimates for the periodic solution of the equation

$$y' = 2y^2 - \cos^2 5x,$$

that lies in the region $y < 0$.

Section 19

NON-LINEAR SYSTEMS

By an appropriate elimination process, a system of differential equations can be reduced to a single equation in one
unknown or to several equations with one unknown function in
each equation. This is a purely algebraic process, at least
in principle.

Example 1. Solve the system of equations

$$y' = \frac{z}{x}, \ z' = \frac{(y - z)^2 + xz}{x^2}. \tag{1}$$

Solution: To eliminate z from the given system we proceed as follows. From the first equation we find $z = xy'$.
Substituting this in the second equation we obtain the following relation.

$$x^3 y'' = (y - xy')^2$$

The given system of equations has been replaced by a single
second order equation. A possible way to solve this equation
is explained in section 10, "Reduction of the Order of an
Equation." Once this equation has been solved for y, the
value of z can be found from the relation $z = xy'$.

If we use this method on most systems of equations the resulting equation in a single unknown usually has very high order. It is therefore more convenient as a rule to attempt to find a combination of the variables which gives an integral of the system. Such a combination is called a first integral.

Example 2. Solve the following system in symmetric form (it is always possible to bring a system into symmetric form by introducing a sufficient number of extra variables)

$$\frac{dx}{xz} = \frac{dy}{yz} = \frac{dz}{-xy}. \tag{2}$$

The first equality of this set is an integrable combination. If we simply multiply both sides of this first equality by z and integrate we obtain the following first integral:

$$\frac{x}{y} = C_1 \tag{3}$$

A second integrable combination is obtained by using the following assertion. If the relations

$$\frac{a_1}{b_1} = \frac{a_2}{b_2} = \ldots = \frac{a_n}{b_n} = t,$$

hold, then the following relations will hold for arbitrary values of $k_1, k_2, \cdots, k_n$:

$$\frac{k_1 a_1 + k_2 a_2 + \ldots + k_n a_n}{k_1 b_1 + k_2 b_2 + \ldots + k_n b_n} = t.$$

This assertion allows us to obtain from (2)

$$\frac{y \cdot dx + x \cdot dy}{y \cdot xz + x \cdot yz} = \frac{dz}{-xy}; \quad \frac{d(xy)}{2xyz} = \frac{dz}{-xy}; \quad d(xy) = -2z\,dz.$$

thus we obtain another integral

$$xy + z^2 = C_2. \tag{4}$$

Obviously the two integrals (3) and (4) are independent and the system has been solved.

It would have been possible to proceed in the following alternative manner. By using the first integral (3) we could eliminate the unknown x in the second equation of (2). Indeed, if we substitute $x = C_1 y$ in this second equation, we obtain

$$\frac{dy}{yz} = \frac{dz}{-C_1 y^2} \ .$$

But this is an integrable equation, indeed the variables are separable: $-C_1 y\,dy = z\,dz$; $z^2 = -C_1 y^2$.

The last step is to replace C_1 by its value from formula (3), and thus we get another first integral:

$$z^2 + xy = C_2$$

Solve the given systems of equations 921-940.

921. $y' = \dfrac{x}{z}$, $z' = -\dfrac{x}{y}$. **922.** $y' = \dfrac{y^2}{z-x}$, $z' = y+1$.

923. $y' = \dfrac{z}{x}$, $z' = \dfrac{z(y+2z-1)}{x(y-1)}$. **924.** $y' = y^2 z$, $z' = \dfrac{z}{x} - yz^2$.

925. $2zy' = y^2 - z^2 + 1$, $z' = z + y$. **926.** $\dfrac{dx}{2y-z} = \dfrac{dy}{y} = \dfrac{dz}{z}$.

927. $\dfrac{dx}{y} = \dfrac{dy}{x} = \dfrac{dz}{z}$. **928.** $\dfrac{dx}{y+z} = \dfrac{dy}{x+z} = \dfrac{dz}{x+y}$.

929. $\dfrac{dx}{y-x} = \dfrac{dy}{x+y+z} = \dfrac{dz}{x-y}$. **930.** $\dfrac{dx}{z} = \dfrac{dy}{u} = \dfrac{dz}{x} = \dfrac{du}{y}$.

931. $\dfrac{dx}{y-u} = \dfrac{dy}{z-x} = \dfrac{dz}{u-y} = \dfrac{du}{x-z}$.

932. $\dfrac{dx}{z} = \dfrac{dy}{xz} = \dfrac{dz}{y}$. **933.** $\dfrac{dx}{z^2 - y^2} = \dfrac{dy}{z} = -\dfrac{dz}{y}$.

934. $\dfrac{dx}{x} = \dfrac{dy}{y} = \dfrac{dz}{xy+z}$. **935.** $\dfrac{dx}{xz} = \dfrac{dy}{yz} = \dfrac{dz}{xy\sqrt{z^2+1}}$.

936. $\dfrac{dx}{x+y^2+z^2} = \dfrac{dy}{y} = \dfrac{dz}{z}$.

937. $\dfrac{dx}{x(y+z)} = \dfrac{dy}{z(z-y)} = \dfrac{dz}{y(y-z)}$.

938. $-\dfrac{dx}{x^2} = \dfrac{dy}{xy-2z^2} = \dfrac{dz}{xz}$.

939. $\dfrac{dx}{x(z-y)} = \dfrac{dy}{y(y-x)} = \dfrac{dz}{y^2-xz}$.

940. $\dfrac{dx}{x(y^2-z^2)} = -\dfrac{dy}{y(z^2+x^2)} = \dfrac{dz}{z(x^2+y^2)}$.

In problems 941-943 the functions ϕ are testing functions from which you must determine whether $\phi = C$ is a first integral of the system beside which it stands.

941. $\dfrac{dx}{dt} = \dfrac{x^2-t}{y}$, $\dfrac{dy}{dt} = -x$; $\varphi_1 = t^2 + 2xy$; $\varphi_2 = x^2 - ty$.

942. $\dot{x} = xy$, $\dot{y} = x^2 + y^2$; $\varphi_1 = x \ln y - x^2 y$; $\varphi_2 = \dfrac{y^2}{x^2} - 2 \ln x$.

943. $\dfrac{dx}{y} = -\dfrac{dy}{x} = \dfrac{dz}{u} = -\dfrac{du}{z}$; $\varphi = yz - ux$.

944. Determine whether the two first integrals

$$\frac{x + y}{z + x} = C_1 \, , \, \frac{z - y}{x + y} = C_2$$

of the system

$$\frac{dx}{x} = \frac{dy}{y} = \frac{dz}{z} \, .$$

are independent.

945.[*] Let $\phi(x,y)$ be a continuous function of x and y,

and suppose that the origin is a singular point, in fact is

either a node or a focus of the system

$$dx/dt = P(x,y), \quad dy/dt = Q(x,y).$$

Show that $\phi(x,y) = C$ cannot be a first integral of the system.

Section 20

FIRST ORDER PARTIAL DIFFERENTIAL EQUATIONS

Let $a_1, \ldots, a_n, b$ be functions of $x_1, \ldots x_n, z$

To solve the first order differential equation

$$a_1 \frac{\partial z}{\partial x} + \cdots + a_n \frac{\partial z}{\partial x_n} = b, \qquad (1)$$

It is necessary to solve the following symmetric system of first order differential equations

$$\frac{dx_1}{a_1} = \cdots = \frac{dx_n}{a_n} = \frac{dz}{b} \qquad (2)$$

indeed to find n independent first integrals of this system:

$$\phi_1 (x_1, \ldots, x_n, z) = C_1$$

$$\phi_n (x_1, \ldots, x_n, z) = C_n. \qquad (3)$$

If these integrals can be found, then every solution of system (1) can be written in the form

$$F (\phi_1, \ldots, \phi_n) = 0, \qquad (4)$$

where F is an arbitrary differential function.

In particular, suppose z occurs explicitly in just one of the first integrals (3) for example, in the last one. Then

if F is an arbitrary differentiable function the general solution can be written in the form

$$\phi_n (x_1, \ldots, x_n. z) = f (\phi_1, \ldots, \phi_{n-1}), \tag{5}$$

If we then solve (5) for z, we obtain the general solution of
(1) in explicit form.

Suppose we wish to find a surface $z = z(x,y)$ that satisfies the differential equation

$$a_1 (x, y, z) \frac{\partial z}{\partial x} + a_2 (x, y, z) \frac{\partial z}{\partial y} = b (x, y, z) \tag{6}$$

and that goes through a given curve

$$x = u(t), \quad y = v(t), z = w(t). \tag{7}$$

The following procedure will be successful. We first find two
independent first integrals of the system

$$\frac{dx}{a_1} = \frac{dy}{a_2} = \frac{dz}{b} . \tag{8}$$

Suppose these are

$$\phi_1 (x, y, z) = C_1, \quad \phi_2 (x, y, z) = C_2 \tag{9}$$

In these latter equations we replace x,y,z by their values from
equation (7) and obtain in this way two equations in the parameter t:

$$\vartheta_1 (t) = C_1, \quad \vartheta_2 (t) = C_2 \tag{10}$$

We now eliminate T and obtain a relation of the following form:

$$F \ (C_1 \ , \ C_2 \ ) = 0.$$

Finally we replace C_1 , C_2 by the functions ϕ_1 , ϕ_2 from the left members of the integrals (9) and have a surface of the required type.

If it happens that the two equations (10) are independent of t, it must be true that the curve (7) lies on every solution of the system (6), that is, it is a characteristic of this system. In this case the Cauchy problem has infinitely many solutions.

Example. Find the general solution of the equation

$$xz \ \partial z/\partial \ x + yz \ \partial z/\partial y = -xy \qquad (11)$$

and find a surface which contains the curve

$$y = x^2, \ z = x^3 \qquad (12)$$

Solution: The system of equations to be solved is

$$\frac{dx}{xz} = \frac{dy}{yz} = \frac{dz}{-xy}$$

for which two first integrals were found in section 19:

$$\frac{x}{y} = C_1 , \ z^2 + xy = C_2 . \qquad (13)$$

Therefore, the general solution of (11) can be written in the form

$$F\left(\frac{x}{y},\ z^2 + xy\right) = 0,$$

where F is an arbitrary function. Since z is absent from one of the first integrals, the general solution can also be written in explicit form as follows:

$$z + xy = f\left(\frac{x}{y}\right); \quad z = \pm\sqrt{f\left(\frac{x}{y}\right) - xy}$$

Here F is an arbitrary function.

To find the integral surface which contains the curve (12) we must first write the equations of this curve in parametric form:

$$x = x, \quad y = x^2, \quad z = x^3.$$

We now substitute this in (13) and obtain

$$1/x = C_1, \quad x^6 + x^3 = C_2.$$

x can be eliminated as follows:

$$\frac{1}{C_1^6} + \frac{1}{C_1^3} = C_2.$$

In this latter equation C_1 and C_2 stand for expressions in the system (13) and thus the integral surface required is:

$$\left(\frac{y}{x}\right)^6 + \left(\frac{y}{x}\right)^3 = z^2 + xy.$$

124

Find the general solution for problems 946-963.

946. $y \frac{\partial z}{\partial x} + x \frac{\partial z}{\partial y} = x - y.$ **947.** $e^x \frac{\partial z}{\partial x} + y^2 \frac{\partial z}{\partial y} = ye^x.$

948. $2x \frac{\partial z}{\partial x} + (y - x) \frac{dz}{dy} - x^2 = 0.$

949. $xy \frac{\partial z}{\partial x} - x^2 \frac{\partial z}{\partial y} = yz.$

950. $x \frac{\partial z}{\partial x} + 2y \frac{\partial z}{\partial y} = x^2 y + z.$

951. $(x^2 + y^2) \frac{\partial z}{\partial x} + 2xy \frac{\partial z}{\partial y} + z^2 = 0.$

952. $2y^4 \frac{\partial z}{\partial x} - xy \frac{\partial z}{\partial y} = x \sqrt{z^2 + 1}.$

953. $x^2 z \frac{\partial z}{\partial x} + y^2 z \frac{\partial z}{\partial y} = x + y.$

954. $yz \frac{\partial z}{\partial x} - xz \frac{\partial z}{\partial y} = e^z.$

955. $(z - y)^2 \frac{\partial z}{\partial x} + xz \frac{\partial z}{\partial y} = xy.$

956. $xy \frac{\partial z}{\partial x} + (x - 2z) \frac{\partial z}{\partial y} = yz.$

957. $y \frac{\partial z}{\partial x} + z \frac{\partial z}{\partial y} = \frac{y}{x}.$

958. $\sin^2 x \frac{\partial z}{\partial x} + \operatorname{tg} z \frac{\partial z}{\partial y} = \cos^2 z.$

959. $(x + z) \frac{\partial z}{\partial x} + (y + z) \frac{\partial z}{\partial y} = x + y.$

960. $(xz + y) \frac{\partial z}{\partial x} + (x + yz) \frac{\partial z}{\partial y} = 1 - z^2.$

961. $(y + z) \frac{\partial u}{\partial x} + (z + x) \frac{\partial u}{\partial y} + (x + y) \frac{\partial u}{\partial z} = u.$

962. $x \frac{\partial u}{\partial x} + y \frac{\partial u}{\partial y} + (z + u) \frac{\partial u}{\partial z} = xy.$

963. $(u - x) \frac{\partial u}{\partial x} + (u - y) \frac{\partial u}{\partial y} - z \frac{\partial u}{\partial z} = x + y.$

In problems 964-980 find the surface which satisfies the differential equation in question and contains the given curve.

964. $y^2 \frac{\partial z}{\partial x} + xy \frac{\partial z}{\partial y} = x;$ $x = 0,$ $z = y^2.$

965. $x \frac{\partial z}{\partial x} - 2y \frac{\partial z}{\partial y} = x^2 + y^2;$ $y = 1,$ $z = x^2.$

966. $x \frac{\partial z}{\partial x} + y \frac{\partial z}{\partial y} = z - xy;$ $x = 2,$ $z = y^2 + 1.$

967. $\operatorname{tg} x \frac{\partial z}{\partial x} + y \frac{\partial z}{\partial y} = z;$ $y = x,$ $z = x^3.$

968. $x \frac{\partial z}{\partial x} - y \frac{\partial z}{\partial y} = z^2(x - 3y);$ $x = 1,$ $yz + 1 = 0.$

969. $x \frac{\partial z}{\partial x} + y \frac{\partial z}{\partial y} = z - x^2 - y^2;$ $y = -2,$ $z = x - x^2.$

970. $yz \frac{\partial z}{\partial x} + xz \frac{\partial z}{\partial y} = xy;$ $x = a,$ $y^2 + z^2 = a^2.$

971. $z \frac{\partial z}{\partial x} - xy \frac{\partial z}{\partial y} = 2xz;$ $x + y = 2,$ $yz = 1.$

972. $z \frac{\partial z}{\partial x} + (z^2 - x^2) \frac{\partial z}{\partial y} + x = 0;$ $y = x^2,$ $z = 2x.$

973. $(y - z) \frac{\partial z}{\partial x} + (z - x) \frac{\partial z}{\partial y} = x - y;$ $z = y = -x.$

974. $x \frac{\partial z}{\partial x} + (xz + y) \frac{\partial z}{\partial y} = z;$ $x + y = 2z,$ $xz = 1.$

975. $y^2 \frac{\partial z}{\partial x} + yz \frac{\partial z}{\partial y} + z^2 = 0;$ $x - y = 0,$ $x - yz = 1.$

976. $x \frac{\partial z}{\partial x} + z \frac{\partial z}{\partial y} = y;$ $y = 2z,$ $x + 2y = z.$

977. $(y + 2z^2) \frac{\partial z}{\partial x} - 2x^2z \frac{\partial z}{\partial y} = x^2;$ $x = z,$ $y = x^2.$

978. $(x - z) \frac{\partial z}{\partial x} + (y - z) \frac{\partial z}{\partial y} = 2z;$ $x - y = 2,$ $z + 2x = 1.$

979. $xy^3 \frac{\partial z}{\partial x} + x^2z^2 \frac{\partial z}{\partial y} = y^3z;$ $x = -z^3,$ $y = z^2.$

980*. $x \frac{\partial z}{\partial x} + y \frac{\partial z}{\partial y} = 2xy;$ $y = x,$ $z = x^2.$

981. Find the general equation of the surfaces which intersect the surfaces of the family $z^2 = Cxy$ orthogonally.

126

982. Find the equation of a surface which contains the
line

$$y = x, \ z = 1 \quad \text{and is orthogonal to the}$$

surfaces

$$x^2 + y^2 + z^2 = Cx.$$

983. Find the partial differential equation that is satisfied by cylindrical surfaces with elements parallel to the
vector $[1,1,2]$. State the general solution of this equation.

984. Use the result of the preceding problem to find the
equation of a cylindrical surface with elements parallel to the
vector $[1,1,2]$, given that its generating curve is

$$x + y + z = 0, \ 5x^2 + 6xy + 5y^2 = 4.$$

985. Set up and solve the partial differential equation
that is satisfied by all conical surfaces with vertex in a fixed
point (a,b,c).

986. Find the equation of the family of surfaces with the
following property: The abscissa of the point of intersection
of an arbitrary tangent plane with the x-axis is half as great
as the abscissa of the point of tangency.

In problems 987-989 solve the given system of equations.

987. $\begin{cases} \dfrac{\partial z}{\partial x} = \dfrac{y}{x}, \\ \dfrac{\partial z}{\partial y} = \dfrac{2z}{y}. \end{cases}$ 988. $\begin{cases} \dfrac{\partial z}{\partial x} = y - z, \\ \dfrac{\partial z}{\partial y} = xz. \end{cases}$

989. $\begin{cases} \dfrac{\partial z}{\partial x} = 2yz - z^2, \\ \dfrac{\partial z}{\partial y} = xz. \end{cases}$

The problem of Pfaff is to be solved by rewriting the differential equation so that it becomes exact. This leads to partial differential equations for the integrating factor; the latter are to be solved by the methods already explained.

In problems 990-993 find the surface which satisfies the given Pfaffian condition.

990. $(x - y)\,dx + z\,dy - x\,dz = 0.$
991. $3yz\,dx + 2xz\,dy + xy\,dz = 0.$
992. $(z + xy)\,dx - (z + y^2)\,dy + y\,dz = 0.$
993. $(2yz + 3x)\,dx + xz\,dy + xy\,dz = 0.$

ANSWERS

Note. In all places, even where not so indicated, the arguments of logarithms are absolute values. The answer $x = C_1 \exp y$ is sometimes written $y = LN\ Cx$, without any requirement that C and x have the same sign. The precise form would be $y = LN\ |Cx|$. In some cases, printing difficulties have required the omission of the absolute value sign, but it <u>must</u> be understood when needed. Only when the argument of the logarithm is necessarily positive for all allowable real values of the variables can the symbol be omitted.

1. Isoclines, Construction of the Differential Equation for a Family of Curves Isogonal Trajectories

15. $f(x, y) = 0$; $f'_x < 0$ (max), $f'_x > 0$ (min). 16. $f'_x + f \cdot f'_y = 0$.

17. $y = e^{\frac{xy'}{y}}$. 18. $y' = 3y^{\frac{2}{3}}$. 19. $xy' = 3y$. 20. $y^2 + y'^2 = 1$.

21. $x^2 y' - xy = yy'$. 22. $2xyy' - y^2 = 2x^3$. 23. $y'^3 = 4y\,(xy' - 2y)$.

24. $y' = \cos \dfrac{x\sqrt{1 - y'^2}}{y}$. 25. $x\,(x-2)\,y'' - (x^2-2)\,y' + 2\,(x-1)\,y = 0$.

26. $(yy' + y'^2)^2 = -y^3 y''$. 27. $(1 - x\,\operatorname{ctg} x)\,y'' - xy' + y = 0$.

28. $x^3 y''' - 3x^2 y'' + 6xy' - 6y - 0$. 29. $y''' y' = 3y''^2$.

30. $(y - 2x)^2\,(y'^2 + 1) = (2y'^2 + 1)^2$. 31. $xy'^2 = y\,(2y' - 1)$.

32. $[x - y(\sqrt{2} + 1)]^2\,(y'^2 + 1) = (x + yy')^2$. 33. $x^2 y'' - 2xy' + 2y = 0$.

34. $(y''y'^2 + 1)^2 = (y'^2 + 1)^3$. 35. $yy' + zz' = 0$, $y^2 + 2xzz' = x^2 z'^2$.

36. $x^2 + y^2 = z^2 - 2z\,(y - xy')$; $x + yy' = zz' - z'\,(y - xy')$.

37. $4yy' = -x$. 38. $y' = -2y$. 39. $(x^2 + y)\,y' = -x$.

40. $(x + y)\,y' = y - x$; $(x - y)\,y' = x + y$. 41. $(x \mp y\sqrt{3})\,y' = y \pm x\sqrt{3}$. 42. $(3x \mp y\sqrt{3})\,y' = y \pm 3x\sqrt{3}$. 43. $(2x \mp y\sqrt{3})\,y' = y \pm 2x\sqrt{3}$. 44. $r' \sin \theta = r^2$. 45. $r' = \dfrac{1}{2} r\,\operatorname{ctg} \theta$. 46. $r' = r\,\operatorname{ctg}\,(\theta \pm 45°)$.

47. $(x + 2y)\,y' = -3x - y$; $(3x + 2y)\,y' = y - x$. 48. $y'\,[2xy \pm (x^2 - y^2)] = y^2 - x^2 \pm 2xy$. 49. $x\,(1 + y'^2) = -2yy'$. 50. $yy'^3 + xy'^2 = -1$.

2. Equations in which the Variables are Separable.

51. $y = C(x+1)e^{-x};\ x = -1.$ **52.** $\ln|x| = C + \sqrt{y^2+1}.$
53. $y(\ln|x^2-1|+C) = 1,\qquad y = 0;\qquad y[\ln(1-x^2)+1] = 1.$
54. $y = 2 + C\cos x;\ y = 2 - 3\cos x.$ **55.** $y = (x-C)^3;\ y = 0;$
$y = (x-2)^3;\ y = 0.$ **56.** $y(1-Cx) = 1;\ y = 0;\ y(1+x) = 1.$
57. $y^2 - 2 = Ce^{\frac{1}{x}}.$ **58.** $(Ce^{-x^2}-1)y = 2;\ y = 0.$ **59.** $e^{-s} = 1 + Ce^t.$
60. $z = -\lg(C - 10^x).$ **61.** $x^2 + t^2 - 2t = C.$ **62.** $\operatorname{ctg}\dfrac{y-x}{2} = x + C;$
$y - x = 2\pi k,\ k = 0,\ \pm 1,\ \pm 2,\ \dots.$ **63.** $2x + y - 1 = Ce^x.$
64. $x + 2y + 2 = Ce^y;\ x + 2y + 2 = 0.$ **65.** $\sqrt{4x+2y-1}-$
$-2\ln(\sqrt{4x+2y-1}+2) = x + C.$ **66.** $y = \arctg\left(1-\dfrac{2}{x}\right)+2\pi.$
67. $y = 2.$ **68.** a) $2y^2 + x^2 = C;$ b) $y^2 + 2x = C;$ c) $y^2 = Ce^{x^2+y^2}.$

3. Geometrical and Physical Problems.

71. $(C \pm x)\,y = 2a^2.$ **72.** $b\ln y - y = \pm x + C,\ 0 < y < b.$

73. $a\ln(a \pm \sqrt{a^2 - y^2}) \mp \sqrt{a^2 - y^2} = x + C.$ **74.** $y = Cx^2.$ **75.** $y = Cx^2.$

76. $r(1 \pm \cos\phi) = C.$ **77.** 10 min. **78.** 0.5 kg. **79.** 24 min.

80. 40 min. **81.** 7.8 min. **82.** $b - \dfrac{b-a}{60\,k}(1 - e^{-60k}).$

83. 50 sec. 15 m. **84.** 200 days. **85.** 1575 years. **86.** 975.10^6

year. **87.** 98.1%. **88.** 23 sec. **89.** 1.75 sec., 16.3 m, 2 sec.,

20 m. **90.** 1.87 sec., 16.4 m/sec. **91.** 17.5 min. **92.** 17.3 min.

93. $5(2 + \sqrt{2}) = 17.07$ min. **94.** 27 sec. **95.** 260 sec., 200 sec.

96. 0.5 kPℓ. **97.** $p \approx \exp(-0.12h)$, where p is measured in

kg/cm^2, and h is measured in km. **98.** 5350 kg.

99. $m_0 - v(q_1 - q_0)(1 - e^{-kt})$, where k is the coefficient of

proportionality. **100.** $c\ln(M/m).$

4. Homogeneous Equations.

101. $x + y = Cx^2$; $x = 0$.   **102.** $\ln(x^2 + y^2) = C - 2\,\text{arctg}\,\dfrac{y}{x}$.

103. $x(y - x) = Cy$; $y = 0$. **104.** $x = \pm y\sqrt{\ln Cx}$; $y = 0$. **105.** $y = Ce^{\frac{y}{x}}$.

106. $y^2 - x^2 = Cy$, $y = 0$. **107.** $\sin\dfrac{y}{x} = Cx$. **108.** $y = -x\ln\ln Cx$.

109. $\ln\dfrac{x+y}{x} = Cx$. **110.** $\ln Cx = \text{ctg}\left(\dfrac{1}{2}\ln\dfrac{y}{x}\right)$; $y = xe^{2\pi k}$, $k = 0$,

± 1, ± 2, ... **111.** $x\ln Cx = 2\sqrt{xy}$; $y = 0$. **112.** $\arcsin\dfrac{y}{x} = \ln Cx \cdot \text{sgn}\,x$;

$y = \pm x$. **113.** $(y - 2x)^3 = C(y - x - 1)^2$; $y = x + 1$. **114.** $2x + y - 1 = Ce^{2y - x}$.
115. $(y - x + 2)^2 + 2x = C$. **116.** $(y - x + 5)^5(x + 2y - 2) = C$.

117. $(y + 2)^2 = C(x + y - 1)$; $y = 1 - x$. **118.** $y + 2 = Ce^{-2\,\text{arctg}\,\frac{y+2}{x-3}}$.

119. $\ln\dfrac{y+x}{x+3} = 1 + \dfrac{C}{x+y}$.   **120.** $\sin\dfrac{y - 2x}{x + 1} = C(x + 1)$.

121. $x^2 = (x^2 - y)\ln Cx$; $y = x^2$.   **122.** $x = -y^2\ln Cx$; $y = 0$.

123. $x^2 y^4 \ln Cx^2 = 1$; $y = 0$.   **124.** $y^2 e^{-\frac{1}{xy}} = C$; $y = 0$; $x = 0$.
125. $(2\sqrt{y} - x)\ln C(2\sqrt{y} - x) = x$; $2\sqrt{y} = x$.   **126.** $1 - xy =$
$= Cx^3(2 + xy)$; $xy = -2$. **127.** $2\sqrt{\dfrac{1}{xy^2} - 1} = -\ln Cx$; $xy^2 = 1$.

128. $\arcsin\dfrac{y^2}{|x^3|} = \ln Cx^3$; $|x^3| = y^2$.   **129.** $x^2 y\ln Cy = 1$; $y = 0$.
130. a) $y^2 = C(x + y)$; $y = -x$; b) $(y + x)^2(y - 2x)^4 = C(y - x)^3$;
$y = x$. **131.** $y = C(x^2 + y^2)$. **132.** $x^2 + y^2 = Cx$. **133.** for $\dfrac{1}{\beta} - \dfrac{1}{\alpha} = 1$.

135. $f(t) \neq t$, $f(+\infty) = f(-\infty) \neq \infty$, $\displaystyle\int_{-\infty}^{+\infty} \dfrac{(tf(t) + 1)\,dt}{(f(t) - t)(t^2 + 1)} = 0$.

5. Linear First Order Equations.

136. $y = Cx^2 + x^4$.   **137.** $y = (2x + 1)(C + \ln|2x + 1|) + 1$.
138. $y = \sin x + C\cos x$. **139.** $y = e^x(\ln|x| + C)$. **140.** $xy = C - \ln|x|$.
141. $y = x(C + \sin x)$. **142.** $y = Ce^{x^2} - x^2 - 1$. **143.** $y = C\ln^2 x - \ln x$.
144. $xy = (x^3 + C)e^{-x}$. **145.** $x = y^2 + Cy$; $y = 0$. **146.** $x = e^y + Ce^{-y}$.
147. $x = (C - \cos y)\sin y$.   **148.** $x = 2\ln y - y + 1 + Cy^2$; $y = 0$.
149. $x = Cy^3 + y^2$; $y = 0$. **150.** $(y - 1)^2 x = y - \ln Cy$; $y = 0$; $y = 1$.
151. $y(e^x + Ce^{2x}) = 1$; $y = 0$.   **152.** $y(x + 1)(\ln|x + 1| + C) = 1$;
$y = 0$. **153.** $y^{-3} = C\cos^3 x - 3\sin x\cos^2 x$; $y = 0$. **154.** $y^3 = Cx^3 - 3x^2$.
155. $y^2 = Cx^2 - 2x$; $x = 0$.   **156.** $y = x^4\ln^2 Cx$; $y = 0$.
157. $y^{-2} = x^4(2e^x + C)$; $y = 0$.   **158.** $y^2 = x^2 - 1 + C\sqrt{|x^2 - 1|}$.
159. $x^2(C - \cos y) = y$; $y = 0$.   **160.** $xy(C - \ln^2 y) = 1$.
161. $x^2 = Ce^{2y} + 2y$. **162.** $y^2 = C(x + 1)^2 - 2(x + 1)$. **163.** $e^{-y} = Cx^2 + x$.
164. $\cos y = (x^2 - 1)\ln C(x^2 - 1)$. **165.** $y = 2e^x - 1$. **166.** $y = -2e^x$.

167. $y = \dfrac{2}{x} + \dfrac{4}{Cx^5 - x}$; $y = \dfrac{2}{x}$. **168.** $y = \dfrac{1}{x} + \dfrac{1}{\dfrac{2}{Cx^3} + x}$; $y = \dfrac{1}{x}$.

169. $y = x + \dfrac{x}{x + C}$; $y = x$. **170.** $y = x + 2 + \dfrac{4}{Ce^{4x} - 1}$; $y = x + 2$.

171. $y = e^x - \dfrac{1}{x + C}$; $y = e^x$.   **172.** $3x = C\sqrt{|y|} - y^2$; $y = 0$.

173. $xy = Cx^3 + 2a^2$. 74. $xy = a^2 + Cy^2$. 175. In 20 min., 3.68

kg. 176. In 62 days. 177. In 24 days; in 23 year.

178. $y = \operatorname{tg} x - \sec x$. 179. $x(t) = \int\limits_{-\infty}^{t} e^{s-t} f(s)\, ds;\quad |x(t)| \leqslant M.$

182. $\dfrac{b}{a}$. 183. $\dfrac{b}{a}$. 184. $y(x) = -\int\limits_{0}^{\infty} \sin(x+s) e^{-\frac{s}{2} - \frac{1}{2}\sin s \cdot \cos(s+2x)}\, ds.$

185. $y(x) = x \int\limits_{+\infty}^{x} e^{x^2 - t^2}\, dt \to -\dfrac{1}{2}\, \text{for } x \to +\infty.$

6. Exact Equations.

186. $3x^2 y - y^3 = C$. 187. $x^2 - 3x^3 y^2 + y^4 = C$. 188. $xe^{-y} - y^2 = C$.

189. $4y\ln x + y^4 = C$. 190. $x + \dfrac{x^3}{y^2} + \dfrac{5}{y} = C$. 191. $x^2 + \dfrac{2}{3}(x^2 - y)^{\frac{3}{2}} = C$.

192. $x - y^2 \cos^2 x = C$. 193. $x^3 + x^3 \ln y - y^2 = C$. 194. $x^2 + 1 =$
$= 2(C - 2x)\sin y$. 195. $2x + \ln(x^2 + y^2) = C$. 196. $x + \operatorname{arctg} \dfrac{x}{y} = C$.

197. $\sqrt{1 + x^2} = xy + C$. 198. $2x^3 y^3 - 3x^2 = C$. 199. $y^2 = x^2(C - 2y)$;
$x = 0$. 200. $(x^2 - C) y = 2x$. 201. $x^2 + \ln y = Cx^3$; $x = 0$.

202. $y\sin xy = C$. 203. $\dfrac{x^2}{2} + xy + \ln|y| = C$; $y = 0$. 204. $-x + 1 =$
$= xy(\operatorname{arctg} y + C)$; $x = 0$; $y = 0$. 205. $x + 2\ln|x| + \dfrac{3}{2} y^2 - \dfrac{y}{x} = C$;

$x = 0$. 206. $\sin \dfrac{y}{x} = Ce^{-x^2}$. 207. $\ln|y| - ye^{-x} = C$; $y = 0$.

208. $\ln\left(\dfrac{x^2}{y^2} + 1\right) = 2y + C$; $y = 0$. 209. $x^2 y \ln Cxy = -1$; $x = 0$;

$y = 0$. 210. $x^2 + y^2 = y + Cx$; $x = 0$. 211. $x^2 y + \ln\left|\dfrac{x}{y}\right| = C$;

$x = 0$; $y = 0$. 212. $2xy^2 + \dfrac{1}{xy} = C$; $x = 0$; $y = 0$.

213. $\ln \dfrac{x+y}{y} + \dfrac{y(1+x)}{x+y} = C$; $y = 0$. 214. $\sin^2 y = Cx - x^2$;
$x = 0$. 215. $y = C\ln x^2 y$. 216. $\sin y = -(x^2 + 1)\ln C(x^2 + 1)$.
217. $xy(C - x^2 - y^2) = 1$; $x = 0$; $y = 0$. 218. $y^2 = Cx^2 e^{x^2 y^2}$.

219. $x\sqrt{1 + \dfrac{y^2}{x^2}} + \ln\left(\dfrac{y}{x} + \sqrt{1 + \dfrac{y^2}{x^2}}\right) = C$; $x = 0$. 220. $x^3 - 4y^2 =$
$= Cy\sqrt[3]{xy}$; $x = 0$; $y = 0$.

8. Equations in which the Derivative Appears Implicitly.

241. $y = Ce^{\pm x}$. 242. $y^2 = (x+C)^3$; $y=0$. 243. $y + x = (x+C)^3$; $y = -x$. 244. $(x+C)^2 + y^2 = 1$; $y = \pm 1$. 245. $y(x+C)^2 = 1$; $y=0$. 246. $y[1+(x-C)^2] = 1$; $y=0$; $y=1$. 247. $(y-x)^2 = 2C(x+y) - C^2$.

$y=0$. 248. $(x-1)^{\frac{4}{3}} + y^{\frac{4}{3}} = C$. 249. $4y = (x+C)^2$; $y = Ce^x$.
250. $y^2(1-y) = (x+C)^2$; $y=1$. 251. $y = Ce^x$; $y = Ce^{-x} + x - 1$.
252. $x^2 y = C$; $y = Cx$. 253. $x^2 + C^2 = 2Cy$; $y = \pm x$. 254. $(x+C)^2 = 4Cy$;
$y=0$; $y=x$. 255. $\ln|1 \pm 2\sqrt{2y-x}| = 2(x + C \pm \sqrt{2y-x})$.

256. $4e^{-\frac{y}{3}} = (x+2)^{\frac{4}{3}} + C$. 257. $y = 2x^2 + C$; $y = -x^2 + C$.

258. $y = Cx^{-3} \pm 2\sqrt{x}$. 259. $\ln Cy = x \pm 2e^{\frac{x}{2}}$, $y = 0$.

260. $\ln Cy = x \pm \sin x$; $y=0$. 261. $\operatorname{arctg} u + \frac{1}{2}\ln\left|\frac{u-1}{u+1}\right| = \pm x + C$,

where $u = \sqrt[4]{1 - \dfrac{1}{y^2}}$; $y=0$; $y = \pm 1$. 262. $x^2 + (Cy+1)^2 = 1$; $y=0$.
263. $(Cx+1)^2 = 1 - y^2$; $y = \pm 1$. 264. $2(x-C)^2 + 2y^2 = C^2$; $y = \pm x$.
265. $y = Ce^{\pm x} - x^2$. 266. $y^2 = C^2 x - C$; $4xy^2 = -1$. 267. $x = p^3 + p$,
$4y = 3p^4 + 2p^2 + C$. 268. $x = \dfrac{2p}{p^2-1}$, $y = \dfrac{2}{p^2-1} - \ln|p^2-1| + C$.

269. $x = p\sqrt{p^2+1}$, $3y = (2p^2-1)\sqrt{p^2+1} + C$. 270. $x = \ln p + \dfrac{1}{p}$;
$y = p - \ln p + C$. 271. $x = 3p^2 + 2p + C$, $y = 2p^3 + p^2$; $y=0$.
272. $x = 2\operatorname{arctg} p + C$, $y = \ln(1+p^2)$; $y = 0$. 273. $x = \ln|p| \pm$
$\pm \dfrac{3}{2}\ln\left|\dfrac{\sqrt{p+1}-1}{\sqrt{p+1}+1}\right| \pm 3\sqrt{p+1} + C$, $y = p \pm (p+1)^{\frac{3}{2}}$, $y = \pm 1$.
274. $x = e^p + C$, $y = (p-1)e^p$; $y = -1$.
275. $x = \pm\left(2\sqrt{p^2-1} + \arcsin\dfrac{1}{|p|}\right) + C$, $y = \pm p\sqrt{p^2-1}$; $y=0$.

276. $x = \pm\left(\ln\left|\dfrac{1-\sqrt{1-p}}{1+\sqrt{1-p}}\right| + 3\sqrt{1-p}\right) + C$, $y = \pm p\sqrt{1-p}$;
$y=0$. 277. $x = \pm 2\sqrt{1+p^2} - \ln(\sqrt{p^2+1}\pm 1) + C$,
$y = -p \pm p\sqrt{p^2+1}$; $y=0$. 278. $4y = C^2 - 2(x-C)^2$; $2y = x^2$.
279. $x = -\dfrac{p}{2} + C$, $5y = C^2 - \dfrac{5p^2}{4}$; $x^2 = 4y$. 280. $\pm xp\sqrt{2\ln Cp} = 1$,
$y = \mp\left(\sqrt{2\ln Cp} - \dfrac{1}{\sqrt{2\ln Cp}}\right)$. 281. $pxy = y^2 + p^3$; $y^2(2p+C) = p^4$;
$y=0$. 282. $y^2 = 2Cx - C\ln C$; $2x = 1 + 2\ln|y|$. 283. $Cx = \ln Cy$;
$y = ex$. 284. $xp^2 = C\sqrt{|p|} - 1$, $y = xp - x^2p^3$; $y=0$.
285. $2p^2 x = C - C^2 p^2$, $py = C$; $32x^3 = -27y^4$. 286. $y^2 = 2C^3 x + C^2$;
$27x^2 y^2 = 1$. 287. $y = Cx - C^2$; $4y = x^2$. 288. $x\sqrt{p} = \ln p + C$,
$y = \sqrt{p}(4 - \ln p - C)$; $y=0$. 289. $x = 3p^2 + C|p|^{-\frac{3}{2}}$,
$y = 2p^3 + 3C|p|^{-\frac{1}{2}}\operatorname{sgn} p$; $y=0$. 290. $y = Cx - C - 2$. 291. $x = Cp$,
$2y = C(p^2+1)$; $y = \pm x$. 292. $x = C(p-1)^{-2} + 2p + 1$,
$y = Cp^2(p-1)^{-2} + p^2$; $y=0$; $y=x-2$. 293. $y = Cx - \ln C$;
$y = \ln x + 1$. 294. $y = 2\sqrt{Cx} + C$; $y = -x$. 295. $2C^2(y - Cx) = 1$;
$8y^3 = 27x^2$. 296. $xp^2 = p + C$, $y = 2 + 2Cp^{-1} - \ln p$.
297. $C^3 = 3(Cx - y)$; $9y^2 = 4x^3$. 298. $xy = a^2$. 299. $x^2 + y^2 = 1$.
300. $x = \dfrac{p(p^2+2)}{\sqrt{p^2+1}^3}$, $y = \dfrac{p^2}{\sqrt{p^2+1}^3}$; $x = \dfrac{p}{\sqrt{p^2+1}^3}$, $y = \dfrac{2p^2+1}{\sqrt{p^2+1}^3}$.

9. Miscellaneous First Order Equations.

301. $y = x(Ce^{-x} - 1)$. **302.** $(Cx+1)y = (Cx-1)$; $y=1$.
303. $y(x^2 - C) = x$; $y=0$. **304.** $x(C-y) = C^2$; $x=4y$.
305. $y(x+C) = x+1$; $y=0$. **306.** $x = Cy + y^3$; $y=0$. **307.** $y=C$; $y = C \pm e^x$. **308.** $y\ln Cx = -x$; $y=0$. **309.** $y^2 = C(x^2-1)$; $x=\pm1$.
310. $2y = 2C(x-1) + C^2$; $2y = -(x-1)^2$. **311.** $x = Cy + \ln^2 y$.

312. $y = Cx^2 e^{-\frac{3}{x}}$. **313.** $(x-C)^2 + y^2 = C$; $4(y^2 - x) = 1$.
314. $4x^2 y = (x+2C)^2$; $y=0$. **315.** $x = Ce^y + y^2 + 2y + 2$.
316. $3y = 3C(x-2) + C^3$; $9y^2 = 4(2-x)^3$. **317.** $y^2 = C(xy-1)$; $xy=1$. **318.** $4(x-C)^3 = 27(y-C)^2$; $y=x-1$. **319.** $x+y = \text{tg}(y-C)$.
320. $x^3 y^2 + 7x = C$. **321.** $y(xy-1) = Cx$. **322.** $-e^{-y} = \ln C(x-2)$.

323. $x = y^2(C - 2\ln|y|)$; $y=0$. **324.** $3xy = C \pm 4x^{\frac{3}{2}}$.
325. $y^2(Ce^{x^2} + 1) = 1$; $y=0$. **326.** $y^2 = 2x \ln Cy$; $y=0$.
327. $\ln(x^2 + y^2) + \text{arctg}\frac{y}{x} = C$. **328.** $(x-1)^2 y = x - \ln|x| + C$.

329. $C^2 x^2 + 2y^2 = 2C$; $2x^2 y^2 = 1$. **330.** $y(C\sqrt{|x^2-1|} - 2) = 1$; $y=0$.
331. $y^2(Ce^{2x} + x + 0{,}5) = 1$; $y=0$. **332.** $y^2 = 1 + C(x+1)^2 e^{-2x}$; $x=-1$. **333.** $y \sin x - \frac{x^3}{3} + \frac{y^2}{2} = C$. **334.** $x = 3p^2 + p^{-1}$, $y = 2p^3 - \ln|p| + C$. **335.** $3y^2 = 2\sin x + C\sin^{-2} x$.
336. $x(e^y + xy) = C$. **337.** $x(p-1)^2 = \ln Cp - p$, $y = xp^2 + p$; $y=0$; $y = x+1$. **338.** $(x+1)y = x^2 + x\ln Cx$. **339.** $y^2 + \sqrt{x^4 + y^4} = C$.
340. $px = C\sqrt{p} - 1$, $y = \ln p - C\sqrt{p} + 1$. **341.** $y = x\,\text{tg}\ln Cx$; $x=0$.

342. $y^{\frac{2}{3}} = Ce^{2x} + \frac{x}{3} + \frac{1}{6}$; $y=0$. **343.** $x = Ce^{\sin y} - 2(1 + \sin y)$.

344. $Cy = C^2 e^x + 1$; $y = \pm 2e^{\frac{x}{2}}$. **345.** $y^2 = (x^2 + C)e^{2x}$.

346. $y = Cx - \sqrt[3]{C^3 - 1}$; $y^{\frac{3}{2}} = x^{\frac{3}{2}} - 1$. **347.** $x(y^2-1)^2 = y^3 - 3y + C$; $y = \pm1$. **348.** $\sqrt{y-x} - \sqrt{x} = C$; $y=x$. **349.** $x\sqrt{y} = \sin x + C$; $y=0$. **350.** $x = 4p^3 - \ln Cp$, $y = 3p^4 - p$; $y=0$.
351. $y^2 + 2x^2 \ln Cy = 0$; $y=0$. **352.** $4x + y - 3 = 2\,\text{tg}(2x + C)$.
353. $xy\cos x - y^2 = C$. **354.** $4Cxy = C^2 x^4 - 1$. **355.** $xy(\ln^2 x + C) = 1$.
356. $2\sqrt{y-x^2} = x\ln Cx$; $y=x^2$. **357.** $\frac{y^2}{2} - \frac{1}{x} - xy = C$; $x=0$.

358. $x = Cy^2 - y^2(y+1)e^{-y}$; $y=0$. **359.** $y(\ln y - \ln x - 1) = C$.
360. $x = 2p - \ln p$, $y = p^2 - p + C$. **361.** $2x^5 - x^2 y^2 + y^3 + x = C$.
362. $(y - 4x + 2)^4(2y + 2x - 1) = C$. **363.** $y^3 = (C - x^3)\sin^3 x$.
364. $p^2 x = p\sin p + \cos p + C$, $py = p\sin p + 2\cos p + 2C$; $y=0$.
365. $x^2 y^2 - 1 = xy\ln Cy^2$; $y=0$. **366.** $y = C\cos x + \sin x$.
367. $|x| = \ln\left(\frac{y}{x} + \sqrt{1 + \frac{y^2}{x^2}}\right) + C$; $x=0$. **368.** $(y-x)^2 =$
$= 2C(x+y) - C^2$; $y^{\frac{2}{3}} - x^{\frac{2}{3}} = C$; $y=0$. **369.** $27(y-2x)^2 = (C-2x)^3$;

$y = 2x.$ **370.** $\sin \dfrac{y}{x} = -\ln Cx.$ **371.** $x^2(x^2y + \sqrt{1+x^4y^2}) = C.$

372. $3\sqrt{y} = x^2 - 1 + C\sqrt[4]{|x^3-1|};\ y=0.$ **373.** $x = \dfrac{C}{p^2} - p - \dfrac{3}{2},$ $y = C\left(\dfrac{2}{p} - 1\right) - \dfrac{p^2}{2};\ y = x+2;\ y=0.$ **374.** $(2x+3y-7)^3 = Ce^{x+2y}.$

375. $(x^2 + y + \ln Cy)\,y = x;\qquad y = 0.$ **376.** $x = 2\sqrt{p^2+1} -$ $- \ln(1+\sqrt{p^2+1}) + \ln Cp,\ y = p\sqrt{p^2+1};\ y=0.$ **377.** $y^2 = C\ln^2 x + 2\ln x.$

378. $x = Cue^u,\ 4y = C^2e^{2u}(2u^2+2u+1);\ x^2 = 2y.$ **379.** $xy^2 \ln Cxy = 1;$ $x=0;\ y=0.$ **380.** $x^2 \sin^2 y = 2\sin^3 y + C.$ **381.** $1 - xy = (Cx - 1)^2;$ $xy = 1.$ **382.** $xe^y = e^x + C.$ **383.** $\sin(y-2x) - 2\cos(y-2x) = Ce^{x+2y}.$

384. $y = (2x + C)\sqrt{x^2+1} - x^2 - Cx - 2.$ **385.** $(y+x^2)^2(2y-x^2) = C.$

386. $(x-1)^2 = y^2(2x - 2\ln Cx);\ y=0.$ **387.** $x = p[\ln(1+\sqrt{p^2+1}) - \ln Cp],$ $2y = xp - \sqrt{p^2+1};\ 2y = -1.$ **388.** $(y+3x+7)(y-x-1)^3 = C.$

389. $\sin y = Ce^{-x} + x - 1.$ **390.** $y = C^2(x-C)^2;\qquad 16y = x^4.$

391. $y^2 = x \diagdown (x+1)\ln C(x+1).$ **392.** $e^y = x^2 \ln Cx.$

393. $(y - 2x\sqrt{y-x^2})(2\sqrt{y-x^2} + x) = C.$ **394.** $xy^2 = \ln x^2 - \ln Cy;$ $x=0,\ y=0.$ **395.** $x(y^2+x^2)^3 = \dfrac{2}{5}y^5 + \dfrac{4}{3}x^3y^3 + 2x^4y + Cx^5;$

$x = 0.$ **396.** $(u-1)\ln Cx^6(u-1)^5(u+2)^4 = 3,\ \text{with}\ u^3 = \dfrac{y^2}{x^2} - 2;$ $y^2 = 3x^2.$ **397.** $\sqrt{y} = (x^2-1)(2\ln|x^2-1| + C);\qquad y = 0.$

398. $x^2 - (x-1)\ln(y+1) - y = C.$ **399.** $\operatorname{tg} y = x^2 + Cx;$ $y = (2k+1)\cdot\dfrac{\pi}{2},\quad k = 0,\ \pm1,\ \pm2,\ \ldots$ **400.** $y^2 = Cx^2 + C^2.$

401. $x^3 = Ce^y - y - 2.$ **402.** $y + 1 = x\ln C(y+1);\qquad y = -1.$

403. $y^2 = 2C^2(x - C);\ 8x^3 = 27y^2.$ **404.** $x^6 = y^3(C - y\ln y + y);$ $y = 0.$ **405.** $\ln C(u-v)^3\left(u^2 + uv + \dfrac{v^2}{3}\right)^2 = 2\operatorname{arctg}\dfrac{2u+v}{v},\text{where } u^3 = y$ $v^3 = x;\ y^2 = x^3.$ **406.** $(y-1)^2 = x^2 + Cx.$ **407.** $(x^2 + y^2)(Cx + 1) = x.$

408. $3x + y^3 - 1 = \operatorname{tg}(3x + C).$ **409.** $(C - x^2)\sqrt{y^2 + 1} = 2x.$

410. $(x^2 + y^2 + 1)^2 = 4x^2 + C.$ **411.** $xy - x^2 = y(y-x)\ln\left|\dfrac{Cy}{y-x}\right|;$ $x = 0;\quad y = 0;\quad y = x.$ **412.** $y = \pm x\operatorname{ch}(x + C);\quad y = \pm x.$

413. $\sqrt{y^2+1} = x(Ce^x - 1).$ **414.** $(y-x)\ln C\dfrac{x-1}{x+1} = 2;\ y = x.$

415. $(Ce^x + 2x + 2)\cos y = 1.$ **416.** $(y^2 - Cx^2 + 1)^2 = 4(1-C)y^2;$

$y = \pm x.$ **417.** $y^2 + xy - 1 = Ce^{\frac{x^2}{2}}.$ **418.** $6x^3y^4 + 2x^3y^3 + 3x^2y^4 = C.$

419. $x + \dfrac{1}{x} + y^2 - 2y + 2 = Ce^{-y};\ x = 0.$ **420.** $e^y(C^2x^2 + 1) = 2C,$ $x^2 = e^{-2y}.$

10. Equations which can be Reduced to Equations of Lower Order.

421. $C_1 x - C_1^2 y = \ln|C_1 x + 1| + C_2;$ $\quad 2y = x^2 + C;$ $\quad y = C.$
422. $9C_1^2(y - C_2)^2 = 4(C_1 x + 1)^3;$ $\quad y = \pm x + C.$ $\quad$ **423.** $C_1 y^2 - 1 =$
$= (C_1 x + C_2)^2.$ **424.** $y^3 = C_1(x + C_2)^2;$ $y = C.$ **425.** $y = C_1 \operatorname{tg}(C_1 x + C_2);$
$\ln\left|\dfrac{y - C_1}{y + C_1}\right| = 2C_1 x + C_2;$ $y(C - x) = 1;$ $y = C.$ **423.** $C_1 y = \pm \sin(C_1 x + C_2);$
$C_1 y = \pm \operatorname{sh}(C_1 x + C_2);$ $\quad y = C \pm x.$ $\quad$ **427.** $y = C_1(x - e^{-x}) + C_2.$
428. $y = C_3 - (x + C_1)\ln C_2(x + C_1);$ $\hspace{2cm} y = C_1 x + C_2.$
429. $y + C_1 \ln|y| = x + C_2;$ $y = C.$ **430.** $2y = C_1 \cos 2x + x^2 + C_2 x + C_3.$
431. $y = C_1[1 \pm \operatorname{ch}(x + C_2)].$ $\quad$ **432.** $x = C_1 p + 3p^2;$ $\quad y = \dfrac{12}{5} p^5 +$
$+ \dfrac{5}{4} C_1 p^4 + C_1^2 \dfrac{p^3}{6} + C_2;$ $\quad y = C.$ $\quad$ **433.** $y = C_1 \dfrac{x^2}{2} - C_1^2 x + C_2;$
$y = \dfrac{x^3}{12} + C.$ $\quad$ **434.** $e^y + C_1 = (x + C_2)^2.$ $\quad$ **435.** $2y = C_1 x^2 -$
$- 2C_1^2(x + C_1)\ln(x + C_1) + C_2 x + C_3;$ $\quad 6y = x^3 + C_1 x + C_2.$
436. $y = \operatorname{ch}(x + C_1) + C_2.$ $\quad$ **437.** $e^y \sin^2(C_1 x + C_2) = 2C_1^2;$
$e^y \operatorname{sh}^2(C_1 x + C_2) = 2C_1^2;$ $e^y(x + C)^2 = 2.$ $\quad$ **438.** $y = C_1 \dfrac{x^3}{6} - C_1^3 \dfrac{x^2}{2} +$
$+ C_2 x + C_3;$ $y = \dfrac{8}{315} x^3 \sqrt[3]{3x} + C_1 x + C_2.$ **439.** $3C_1 y = (x - C_1)^3 + C_2;$
$y = C;$ $y = C - 2x^2.$ **440.** $\ln\left|y^2 + C_1 \pm \sqrt{y^4 + 2C_1 y^2 + 1}\right| = 2x + C_2;$
$y = \pm 1.$ $\quad$ **441.** $x = 3C_1 p^2 + \ln C_2 p,$ $\quad y = 2C_1 p^3 + p;$ $\quad y = C.$
442. $x = C_1 e^p - 2p - 2,$ $y = C_1(p - 1)e^p - p^2 + C_2.$ **443.** $12(C_1 y - x) =$
$= C_1^2(x + C_2)^3 + C_3.$ **444.** $y = C_1(x\sqrt{x^2 - 1} - \ln|x + \sqrt{x^2 - 1}|) +$
$+ x^2 + C_2.$ **445.** $\ln y = C_1 \operatorname{tg}(C_1 x + C_2);$ $\ln\left|\dfrac{\ln y - C_1}{\ln y + C_1}\right| = 2C_1 x + C_2;$
$(C - x)\ln y = 1;$ $\quad y = C.$ $\quad$ **446.** $x = u - \ln|1 + u| + C_2,$ where
$u = \pm\sqrt{1 + 4C_1 y};$ $y = C;$ $y = Ce^{-x}.$ **447.** $C_1^2 y = (C_1^2 x^2 + 1)\operatorname{arctg} C_1 x -$
$- C_1 x + C_2,$ $2y = k\pi x^2 + C,$ $k = 0, \pm 1, \pm 2, \dots$ **448.** $x = \ln|p| +$
$+ 2C_1 p - C_2,$ $\hspace{1.5cm} y = p + C_1 p^2 + C_3;$ $\hspace{1cm} y = C_1 x + C_2.$
449. $C_1^2 y + 1 = \pm \operatorname{ch}(C_1 x + C_2);$ $\hspace{1cm} C_1^2 y - 1 = \sin(C_1 x + C_2);$
$2y = (x + C)^2;$ $\quad y = 0.$ $\quad$ **450.** $y = C_2 - \ln\left|\cos\left(\dfrac{x^2}{2} + C_1\right)\right|.$
451. $6y = x^3 \ln|x| + C_1 x^3 + C_2 x^2 + C_3 x + C_4.$

452. $y = x \displaystyle\int_0^x \dfrac{\sin t}{t}\, dt + \cos x + C_1 x + C_2.$

453. $y = C_1\left[x \displaystyle\int_0^x e^{t^2}\, dt - \dfrac{1}{2}(e^{x^2} - 1)\right] + C_2 x + C_3.$

454. $y = \dfrac{x^2}{2} \displaystyle\int_1^x \dfrac{e^t}{t}\,dt - \dfrac{x+1}{2}\,e^x + C_1 x^2 \ln|x| + C_2 x^2 + C_3 x + C_4.$

455. $y = C_2 e^{C_1 x} + C_3 e^{-C_1 x};\ \ y = C_2 \cos C_1 x + C_3 \sin C_1 x;\ \ y = C_1 x + C_2.$
456. $C_1 y = \ln|C_1 x + C_2| + C_3;\ \ y = C_1 x + C_2.$ **457.** $C_1 y - 1 = C_2 e^{C_1 x};$
$y = C - x;\quad y = 0.$ **458.** $y = \pm \sqrt{C_1 x + C_2} + C_3 x + C_4;$
$y = C_1 x^2 + C_2 x + C_3.$ **459.** $y^2 = x^2 + C_1 x + C_2.$
460. $y = e^{\frac{x^2}{2}}\left(C_1 \displaystyle\int e^{-\frac{x^2}{2}}\,dx + C_2\right) - 1.$ **461.** $y = C_1 \operatorname{tg}(C_1 \ln C_2 x);$
$C_2(y + C_1)|x|^{2C_1} = y - C_1;\ y \ln Cx = -1.$ **462.** $y = 4C_1 \operatorname{tg}(C_1 x^2 + C_2);$
$2\ln\left|\dfrac{y - C_1}{y + C_1}\right| = C_1 x^2 + C_2;\ \ y(C - x^2) = 4;\ \ y = C.$ **463.** $y = C_2 e^{Cx^2}.$

464. $\ln C_2 y = 4x^{\frac{5}{2}} + C_1 x.$ **465.** $y = C_2 (x + \sqrt{x^2 + 1})^{C_1}.$
466. $y^2 = C_1 x^3 + C_2.$ **467.** $y = C_2 x e^{-\frac{C_1}{x}}.$ **468.** $y = C_2 |x|^{C_1 - \frac{1}{2}\ln|x|}.$

469. $y = C_2 \left|\dfrac{x}{x+C_1}\right|^{\frac{1}{C_1}};\ \ y = C.$ **470.** $|y|^{C_1^2 + 1} = C_2\left(x - \dfrac{1}{C_1}\right)|x + C_1|^{C_1^2};$
$y = C.$ **471.** $y = C_2 x (\ln C_1 x)^2;\ \ y = Cx.$ **472.** $\ln|y| = \ln|x^2 - 2x + C_1| + $
$+ \displaystyle\int \dfrac{2\,dx}{(x-1)^2 + C_1 - 1} + C_2;\ \ y = C.$ **473.** $4C_1 y^2 = 4x + x(C_1 \ln C_2 x)^2.$

474. $y = -x \ln(C_2 \ln C_1 x);\ \ y = Cx.$ **475.** $\dfrac{y}{x} = C_2 - 3\ln\left|\dfrac{1}{x} - C_1\right|;$
$y = Cx.$ **476.** $x^2 y = C_1 \operatorname{tg}(C_1 \ln C_2 x);\ C_2(x^2 y + C_1)|x|^{2C_1} = x^2 y - C_1;$
$x^2 y \ln Cx = -1.$ **477.** $4(C_1 y - 1) = C_1^2 \ln^2 C_2 x.$ **478.** $Cy = x^{\frac{3}{2}}(C_2 x^C + 2);$

$y = Cx^{\frac{5}{2}}.$ **479.** $2C_2 x^2 y = (C_2 x - C_1)^2 - 1;\quad xy = \pm 1.$
480. $2C_1 C_2 y = C_2^2 |x|^{2+C_1} + |x|^{2-C_1}.$ **501.** $(3 - x) y^5 = 8(x + 2).$

502. $y(x + 2) = -x - 6.$ **503.** $(1 - \ln x)^2 y = x^2.$ **504.** $y = 3\operatorname{th}^2 \dfrac{x\sqrt{3}}{2} - 2.$

505. $\ln\operatorname{tg}\left(\dfrac{y}{2} + \dfrac{\pi}{6}\right) = 2x + 2.$ **506.** a) $4(C_1 y - 1) = C_1^2 (x + C_2)^2;$

b) $y\sqrt{\dfrac{C_1}{y} - 1} + C_1 \arccos\sqrt{\dfrac{y}{C_1}} = C_2 \pm x.$ **507.** $y = C_2 - $
$- k\ln\cos\left(\dfrac{x}{k} + C_1\right).$ **508.** $y = \dfrac{\mathrm{p}}{2\mathrm{T}}\ x^2 + C_1\ x + C_2.$

509. $ay = \cosh(ax + C_1) + C_2,$ where $a = q/T.$

11. Linear Equations with Constant Coefficients.

511. $y=C_1e^x+C_2e^{-2x}$. **512.** $y=C_1e^{-x}+C_2e^{-3x}$. **513.** $y=C_1+C_2e^{2x}$.

514. $y = C_1e^{2x} + C_2e^{\frac{x}{2}}$. **515.** $y = e^{2x}(C_1\cos x + C_2\sin x)$.
516. $y = e^{-x}(C_1\cos 3x + C_2\sin 3x)$. **517.** $y = C_1\cos 2x + C_2\sin 2x$.
518. $y = C_1e^{2x} + e^{-x}(C_2\cos x\sqrt{3} + C_3\sin x\sqrt{3})$. **519.** $y = C_1e^x + C_2e^{-x} + C_3\cos x + C_4\sin x$. **520.** $y = e^x(C_1\cos x + C_2\sin x) + e^{-x}(C_3\cos x + C_4\sin x)$. **521.** $y = e^{\sqrt{3}x}(C_1\cos x + C_2\sin x) + C_3\cos 2x + C_4\sin 2x + e^{-\sqrt{3}x}(C_5\cos x + C_6\sin x)$.

522. $y = e^x(C_1+C_2x)$. **523.** $y = e^{-\frac{x}{2}}(C_1+C_2x)$. **524.** $y = C_1+C_2x + C_3x^2+e^{3x}(C_4+C_5x)$. **525.** $y = C_1+C_2e^x+C_3e^{-x}+C_4e^{3x}+C_6e^{-3x}$.
526. $y = (C_1+C_2x)\cos x+(C_3+C_4x)\sin x$. **527.** $y = e^x(C_1+C_2x+C_3x^2)$.
528. $y = e^x(C_1+C_2x)+C_3e^{-x}$. **529.** $y = C_1e^x+C_2e^{-x}+C_3e^{2x}+C_4e^{-2x}$.
530. $y = C_1+(C_2+C_3x)\cos 2x+(C_4+C_5x)\sin 2x$.
531. $y = e^x(C_1+C_2x)+C_3e^{-2x}$. **532.** $y = C_1\cos x+C_2\sin x+$ $+C_3\cos x\sqrt{3}+C_4\sin x\sqrt{3}$. **533.** $y = C_1e^{-x}+C_2e^{3x}+\frac{1}{5}e^{4x}$.
534. $y = C_1\cos x+C_2\sin x+(2x-2)e^x$. **535.** $y = C_1e^x+C_2e^{-x}+$ $+xe^x+x^2+2$. **536.** $y = C_1e^x+C_2e^{-2x}+\left(\frac{x^2}{2}-\frac{x}{3}\right)e^x$.
537. $y = C_1e^x+C_2e^{2x}+0.1\sin x+0.3\cos x$. **538.** $y = C_1\cos x+$ $+C_2\sin x-2x\cos x$. **539.** $y = C_1e^x+C_2e^{4x}-(2x^2-2x+3)e^{2x}$.
540. $y = C_1e^x+C_2e^{2x}+(0.1x-0.12)\cos x-(0.3x+0.34)\sin x$.
541. $y = C_1e^x+C_2e^{-4x}-\frac{x}{5}e^{-4x}-\left(\frac{x}{6}+\frac{1}{36}\right)e^{-x}$. **542.** $y = C_1e^x +$
$+C_2e^{-3x}+\left(\frac{x^3}{12}-\frac{x^2}{16}+\frac{x}{32}\right)e^x$. **543.** $y = e^{2x}(C_1\cos 2x+C_2\sin 2x)+$
$+0.25e^{2x}+0.1\cos 2x+0.05\sin 2x$. **544.** $y = C_1e^{3x}+C_2e^{-3x}+$
$+e^{3x}\left(\frac{6}{37}\sin x-\frac{1}{37}\cos x\right)$. **545.** $y = (C_1+C_2x+x^3)e^x$.

546. $y = \left(C_1-\frac{x^2}{4}\right)\cos x+\left(C_2+\frac{x}{4}\right)\sin x$. **547.** $y = (C_1+C_2x)e^{-2x}+$
$+\left(\frac{x}{16}-\frac{1}{32}\right)e^{2x}$. **548.** $y = C_1+C_2e^{5x}-0.2x^3-0.12x^2-0.048x+$ $+0.02(\cos 5x-\sin 5x)$. **575.** $y = e^x(x\ln|x|+C_1x+C_2)$.
576. $y = (e^{-x}+e^{-2x})\ln(e^x+1)+C_1e^{-x}+C_2e^{-2x}$.
577. $y = (C_1+\ln|\sin x|)\sin x+(C_2-x)\cos x$. **578.** $y = \sin 2x \times$ $\times\ln|\cos x|-x\cos 2x+C_1\sin 2x+C_2\cos 2x$. **579.** $y = e^{-x}\times$
$\times\left(\frac{4}{5}(x+1)^{\frac{5}{2}}+C_1+C_2x\right)$. **580.** $y = -\frac{1}{x}+C_1e^x+C_2e^{-x}$.

581. $y = 2+e^{-x}$. **582.** $y = (7-3x)e^{x-2}$. **583.** $y = 2\cos x-5\sin x+2e^x$.
584. $y = e^{2x-1}-2e^x+e-1$. **585.** $y = \frac{\text{sh }x}{\text{sh }1}-2x$.
586. $y = 1-\sin x-\cos x$. **587.** No solution. **588.** $y = 2x-\pi+$ $+\pi\cos x+C\sin x$, C — arbitrary. **589.** $y = C_1x^2+C_2x^3$.
590. $y = C_1x^3+C_2x^{-1}$. **591.** $y = x^3(C_1+C_2\ln|x|+C_3\ln^2|x|)$.
592. $y = C_1+C_2\ln|x|+C_3x^3$. **593.** $y = x(C_1+C_2\ln|x|)+2x^3$.
594. $y = C_1\cos(2\ln|x|)+C_2\sin(2\ln|x|)+2x$. **595.** $y = C_1x^2+$
$+\frac{1}{x}\left(C_2-\frac{2}{3}\ln x-\ln^2 x\right)$. **596.** $y = x^2(C_1\cos\ln|x|+C_2\sin\ln|x|+3)$.
597. $y = C_1x^3+C_2x^{-2}+x^3\ln|x|-2x^2$. **598.** $y = C_1x^2+C_2x^{-1}+$ $+0.1\cos\ln x-0.3\sin\ln x$. **599.** $y = (x-2)^2(C_1+C_2\ln|x-2|)+$
$+x-1.5$. **600.** $y = C_1\left(x+\frac{3}{2}\right)+C_2\left|x+\frac{3}{2}\right|^{\frac{3}{2}}+C_3\left|x+\frac{3}{2}\right|^{\frac{1}{2}}$.

12. Linear Equations with Variable Coefficients.

601. No.   602. Yes.   603. No.   604. No.   605. Yes.   606. No.

607. Yes.   608. No.   609. No.   610. Yes.   611. No.   612. Yes.

613. Yes.   614. Yes.   615. No.   616. No.   617. Yes.   618. No.

619. Yes.   620. No.   621 Yes.   622. No.

624. $y'' - y' \operatorname{ctg} x = 0.$ 625. $(x-1)\,y'' - xy' + y = 0.$

626. $y''' - y'' = 0.$ 627. $(2x^2 + 6x - 9)\,y'' - (4x + 6)\,y' + 4y = 0.$

628. $y'' - y = 0.$ 629. $(x^2 - 2x + 2)\,y''' - x^2 y'' + 2xy' - 2y = 0.$

630. $x^2 y'' - 3xy' + 3y = 0.$ 631. $y = C_1 x + C_2 e^{-2x}.$

632. $y = C_1\left(1 + \dfrac{1}{x}\right) + C_2\left(\dfrac{x}{2} + 1 - \dfrac{x+1}{x}\ln|x+1|\right).$

633. $y = e^x(C_1 x^2 + C_2).$ 634. $xy = C_1 e^{-x} + C_2 e^x.$

635. $y = C_1 \operatorname{tg} x + C_2(1 + x \operatorname{tg} x).$ 636. $y = C_1(1 + x\ln|x|) + C_2 x.$

637. $y = C_1(e^x - 1) + \dfrac{C_2}{e^x + 1}.$ 638. $y = C_1 x + C_2(\ln x + 1).$

639. $y = C_1 \sin x + C_2\left(2 - \sin x \cdot \ln \dfrac{1 + \sin x}{1 - \sin x}\right).$

640. $y = \dfrac{x}{(x^2 + 1)^{\frac{3}{2}}}\left[C_1\left(\ln\left(x + \sqrt{x^2 + 1}\right) - \dfrac{\sqrt{x^2 + 1}}{x}\right) + C_2\right].$

641. $y = C_1 e^{2x} + C_2(3x + 1)\,e^{-x}.$ 642. $y = (C_1 + C_2 x)\,e^{-x^2}.$

643. $y = C_1(2x + 1) + C_2 e^{2x}.$ 644. $y = C_1(x + 1) + C_2 x^{-1}.$

645. $y = C_1(x + 2) + C_2 x^2.$ 646. $y = C_1(x^2 + 2) + C_2 x^3.$

647. $y = C_1(x^2 + 1) + C_2[x + (x^2 + 1)\operatorname{arctg} x].$

648. $y = C_1 \sqrt{|x|} + C_2(x - 2).$ 649. $y = C_1 x + C_2 e^x + C_3 e^{-x}.$

650. $y = C_1 x + C_2 x^{-1} + C_3(x\ln|x| + 1).$ 651. $y = C_1 x + C_2 e^x + C_3(x^2 - 1).$

652. $y = C_1(x + 2) + \dfrac{C_2}{x} + \left(\dfrac{x}{2} + 1\right)\ln|x| + \dfrac{3}{2}.$

653. $y = C_1(2x - 1) + C_2 e^{-x} + \dfrac{x^2 + 1}{2}.$ 654. $y = \dfrac{C_1}{x+1} + \dfrac{C_2}{x-1} + x.$

655. $y = C_1(x^2 + 1) + C_2 x^{-1} + 2x.$ 656. $z'' + z = 0.$ 657. $z'' - z = 0.$

658. $z'' = 0.$ 659. $x^2 z'' - 2z = 0.$ 660. $4x^2 z'' + (4x^2 + 1)\,z = 0.$

661. $y''_{tt} - y = 0.$ 662. $y''_{tt} + y = 0.$ 663. $(t^2 - 1)\,y''_{tt} - 2y = 0.$

664. $y''_{tt} + t^2 y = 0.$ 665. $8y''_{tt} + t^2 y = 0.$ 666. $\dfrac{\pi}{\sqrt{m}};$

The number of zeros lies between $\dfrac{(b - a)\sqrt{m}}{\pi} - 1$ (not inclusive)

and $\dfrac{(b - a)\sqrt{m}}{\pi} + 1$ (inclusive).

667. $0.33 < d < 0.5.$ **668.** $15.7 < d < 32.$
669. $0.49 < d < 1.$ **670.** $0.15 < d < 1.2.$ **671.** $15 \leqslant N \leqslant 41.$
677. $u''_{tt} + (\pm 1 + \psi^3 \psi''_{xx}) u = 0, \ t = \int^{\cdot} \frac{dx}{(\psi(x))^2}, \ y = \psi u.$

In problems 680, 681, 684, 685, 686, 688, 690, the second

solution is obtained by replacing the cos function by the

sin function.

680. $y_1 = \frac{1}{\sqrt{x}} \cos \frac{x^2}{2} + O\left(x^{-\frac{5}{2}}\right).$ **681.** $y_1 = e^{-\frac{x}{2}} \cos e^x + O\left(e^{-\frac{3}{2}x}\right).$

682. $y_{1,2} = x^{\frac{1}{4}} e^{\pm i \sqrt{x}} \left(1 + O\left(x^{-\frac{1}{2}}\right)\right).$ **683.** $y_{1,2} = x^{-\frac{1}{4}} e^{\pm \frac{2}{3} x^{\frac{3}{2}}} \times$
$\times \left(1 + O\left(x^{-\frac{3}{2}}\right)\right).$ **684.** $y_1 = x^{-\frac{3}{4}} \cos 2\sqrt{x} + O\left(x^{-\frac{5}{4}}\right).$

685. $y_1 = e^{\frac{(x-1)^2}{2}} \left[(2x)^{-\frac{1}{4}} \cos \frac{(2x)^{\frac{3}{2}}}{3} + O\left(x^{-\frac{7}{4}}\right) \right].$

686. $y_1 = \frac{1}{x} \cos \frac{x^3}{3} + O\left(\frac{1}{x^2}\right).$ **687.** $y_{1,2} = x^{\frac{1 \pm \sqrt{5}}{2}} (1 + O(x^{-2})).$

688. $y_1 = \sqrt{\frac{x}{\ln x}} \left[\cos\left(\frac{1}{2} \ln^2 x - \frac{1}{8} \ln\ln x\right) + O(\ln^{-2} x) \right].$

689. $y_{1,2} = \left[1 \pm \frac{3}{32 x^2} + \frac{105}{2048 x^4} + O(x^{-6}) \right] \frac{e^{\pm x^2}}{\sqrt{2x}}.$

690. $y_1 = x^{\frac{1}{4}} \left(1 + \frac{3}{64 x}\right) \cos\left(2\sqrt{x} + \frac{3}{16\sqrt{x}}\right) + O\left(x^{-\frac{5}{4}}\right).$

13. Series Development of Solutions of Equations.

691. $y = 1 + x + \frac{x^2}{2} + \frac{2x^3}{3} + \frac{7x^4}{12} + \ldots$

692. $y = 1 + x + \frac{x^3}{3} - \frac{x^4}{3} + \ldots$ **693.** $y = \frac{x^2}{2} + \frac{x^3}{6} + \frac{x^4}{6} + \ldots$

694. $y = x + x^2 - \frac{x^3}{6} - \frac{x^4}{4} - \ldots$

695. $y = 1 + 2(x-1) + 4(x-1)^2 + \frac{25}{3}(x-1)^3 + \frac{81}{4}(x-1)^4 + \ldots$

696. $y = 1 + 2x - \frac{x^2}{2} - \frac{x^3}{3} - \frac{x^4}{3} - \ldots$ **697.** $y = 4 - 2x + 2x^2 -$
$- 2x^3 + \frac{19}{6} x^4 + \ldots$ **698.** $R > 0.73.$

699. The error is less than 0.00 024.

700. $y_1 = 1 + \frac{x^4}{3 \cdot 4} + \frac{x^8}{3 \cdot 4 \cdot 7 \cdot 8} + \ldots,\ y_2 = x + \frac{x^5}{4 \cdot 5} + \frac{x^9}{4 \cdot 5 \cdot 8 \cdot 9} + \ldots$

701. $y_1 = 1 + \frac{x^2}{1} + \frac{x^4}{1 \cdot 3} + \frac{x^6}{1 \cdot 3 \cdot 5} + \ldots,\quad y_2 = x + \frac{x^3}{2} + \frac{x^5}{2 \cdot 4} +$

$+ \frac{x^7}{2 \cdot 4 \cdot 6} + \ldots = xe^{\frac{x^2}{2}}.$ **702.** $y_1 = 1 + x^2 + x^4 + \ldots = \frac{1}{1 - x^2},$

$y_2 = x + x^3 + x^5 + \ldots = \frac{x}{1 - x^2}.$ **703.** $y_1 = 1 - \frac{3}{2} x^2 +$

$+ \frac{3 \cdot 5}{2 \cdot 4} x^4 - \ldots = (1 + x^2)^{-\frac{3}{2}},\qquad y_2 = x - \frac{4}{3} x^3 + \frac{4 \cdot 6}{3 \cdot 5} x^5 - \ldots$

704. $y_1 = 1 - \frac{x^2}{2} - \frac{x^3}{2} - \frac{11 x^4}{24} - \ldots,\ y_2 = x + x^2 + \frac{5 x^3}{6} + \frac{3 x^4}{4} + \ldots$

705. $y_1 = 1 + x - x^3 - x^4 + x^6 + x^7 - \ldots = \frac{1}{1 - x + x^2},\ y_2 = xy_1.$

706. $y_1 = 1 - \frac{x^3}{6} - \frac{x^5}{40} + \ldots,\qquad y_2 = x + \frac{x^3}{6} - \frac{x^4}{12} + \ldots$

707. $y_1 = 1 - \frac{x^3}{6} + \frac{x^5}{120} + \ldots,\qquad y_2 = x - \frac{x^4}{12} + \frac{x^6}{180} + \ldots$

708. $y_1 = 1 + \frac{x^2}{2} + \frac{x^3}{12} + \frac{5 x^4}{72} + \ldots,\qquad y_2 = x + \frac{x^3}{6} + \frac{x^4}{24} + \ldots$

709. $y_1 = 1 - \frac{x^3}{6} + \ldots,\ y_2 = x + \frac{x^3}{3} - \frac{x^4}{12} + \ldots,\ y_3 = x^2 + \frac{x^4}{4} - \ldots$

710. $y_1 = 1 - \frac{x^2}{3!} + \frac{x^4}{5!} - \ldots = \frac{\sin x}{x},\ y_2 = \frac{1}{x} - \frac{x}{2!} + \frac{x^3}{4!} - \ldots = \frac{\cos x}{x}.$

711. $y_1 = \frac{1}{x} + 1 + \frac{x}{2!} + \frac{x^2}{3!} + \ldots = \frac{e^x}{x},\qquad y_2 = x^{\frac{1}{2}}\left(1 + \frac{2x}{5} +\right.$

$\left.+ \frac{(2x)^2}{5 \cdot 7} + \frac{(2x)^3}{5 \cdot 7 \cdot 9} + \ldots\right).$ **712.** $y_1 = x^{\frac{1}{3}}\left(1 + \frac{x^2}{5 \cdot 6} + \frac{x^4}{5 \cdot 6 \cdot 11 \cdot 12} + \ldots\right),$

$y_2 = x^{\frac{2}{3}}\left(1 + \frac{x^2}{6 \cdot 7} + \frac{x^4}{6 \cdot 7 \cdot 12 \cdot 13} + \ldots\right).$ **713.** $y_1 = \frac{1}{x} + 1 + \frac{x}{2},$

$y_2 = x^2 + \frac{x^3}{4} + \frac{x^4}{4 \cdot 5} + \frac{x^5}{4 \cdot 5 \cdot 6} + \ldots = 6\left(\frac{e^x - 1}{x} - 1 - \frac{x}{2}\right).$

714. $y_1 = \frac{1}{x^2} - \frac{1}{x} + \frac{1}{2} + \frac{x^2}{8} + \frac{x^3}{40} + \frac{7 x^4}{720} + \ldots,\quad y_2 = x + \frac{x^2}{2} +$

$+ \frac{x^3}{5} + \frac{x^4}{20} + \ldots$ **715.** $y_1 = x + x^2 + \frac{x^3}{2!} + \frac{x^4}{3!} + \ldots = xe^x.$

716. $y_1 = 1 + \frac{x^2}{2^2} + \frac{x^4}{2^2 \cdot 4^2} + \frac{x^6}{2^2 \cdot 4^2 \cdot 6^2} + \ldots$ **717.** $y_2 = \left(1 + \frac{x^2}{2^2} +\right.$

$\left.+ \frac{x^4}{2^2 \cdot 4^2} + \ldots\right)\ln x - \frac{x^2}{4} - \frac{3 x^4}{128} - \ldots$

718. y_1, y_2 are power series in the extended sense, with irrational exponents. **719.** y_1, y_2 are series with complex exponents. **720.** There is no solution in series, even in the extended sense, since the radius of convergence of the putative solution $y = 1 + 1! x + 2! x^2 + 3! x^3 + \ldots$ is 0.

721. $y = \dfrac{2}{\pi} + \dfrac{4}{\pi} \sum\limits_{k=1}^{\infty} \dfrac{1}{16k^4 - 4k^2 + 1} \cdot \left(\cos 2kx - \dfrac{2k}{4k^2 - 1} \sin 2kx \right).$

722. $y = \sum\limits_{k=1}^{\infty} \dfrac{(k^3 + k)\cos kx - \sin kx}{2^k \, [(k^3 + k)^2 + 1]}.$ **723.** $y = \dfrac{1}{z} + \mu \left(z^2 - \dfrac{1}{z^2} \right) +$

$+ \mu^2 \left(-\dfrac{z^5}{7} + \dfrac{2z}{3} - \dfrac{32}{21z^2} + \dfrac{1}{z^3} \right) + \dots, \qquad z = x + 1.$

724. $y = 2\sqrt{x} + 2\mu \left(x^{-\frac{1}{2}} - x^2 \right) + \mu^2 \cdot \left(\dfrac{x^{\frac{7}{2}}}{4} - \dfrac{4}{3} x + \dfrac{25}{12} x^{-\frac{1}{2}} - x^{-\frac{3}{2}} \right) + \dots$

725. $y = 1 + \mu (x^2 - x) - \dfrac{\mu^2}{6} x (x - 1)^3 + \dots$

14. Linear Systems with Constant Coefficients.

726. $x = C_1 e^t + C_2 e^{5t}, \ y = -C_1 e^t + 3C_2 e^{5t}.$ **727.** $x = C_1 e^{-t} + C_2 e^{3t},$
$y = 2C_1 e^{-t} - 2C_2 e^{3t}.$ **728.** $x = 2C_1 e^{3t} - 4C_2 e^{-3t}, \ y = C_1 e^{3t} + C_2 e^{-3t}.$
729. $x = e^{2t}(C_1 \cos t + C_2 \sin t), \ y = e^{2t} [(C_1 + C_2)\cos t + (C_2 - C_1)\sin t].$
730. $x = e^t (C_1 \cos 3t + C_2 \sin 3t), \qquad y = e^t (C_1 \sin 3t - C_2 \cos 3t).$
731. $x = (2C_2 - C_1)\cos 2t - (2C_1 + C_2)\sin 2t, \ y = C_1 \cos 2t + C_2 \sin 2t.$
732. $x = (C_1 + C_2 t) e^{3t}, \ y = (C_1 + C_2 + C_2 t) e^{3t}.$ **733.** $x = (C_1 + C_2 t) e^t,$
$y = (2C_1 - C_2 + 2C_2 t) e^t.$ **734.** $x = (C_1 + 2C_2 t) e^{-t}, \ y = (C_1 + C_2 + 2C_2 t) e^{-t}.$
735. $x = (C_1 + 3C_2 t) e^{2t}, \ y = (C_2 - C_1 - 3C_2 t) e^{2t}.$ **736.** $x = C_1 e^t +$
$+ C_2 e^{2t} + C_3 e^{-t}, \quad y = C_1 e^t - 3C_3 e^{-t}, \quad z = C_1 e^t + C_2 e^{2t} - 5C_3 e^{-t}.$
737. $x = C_1 + 3C_2 e^{2t}, \ y = -2C_2 e^{2t} + C_3 e^{-t}, \ z = C_1 + C_2 e^{2t} - 2C_3 e^{-t}.$
738. $x = C_2 e^{2t} + C_3 e^{3t}, \ y = C_1 e^t + C_2 e^{2t}, \ z = C_1 e^t + C_2 e^{2t} + C_3 e^{3t}.$
739. $x = C_1 e^t + C_2 e^{2t} + C_3 e^{5t}, \qquad y = C_1 e^t - 2C_2 e^{2t} + C_3 e^{5t},$
$z = -C_1 e^t - 3C_2 e^{2t} + 3C_3 e^{5t}.$ **740.** $x = C_1 e^t + C_3 e^{-t}, \ y = C_1 e^t + C_2 e^{2t},$
$z = 2C_2 e^{2t} - C_3 e^{-t}.$ **741.** $x = e^t (2C_2 \sin 2t + 2C_3 \cos 2t),$
$y = e^t (C_1 - C_2 \cos 2t + C_3 \sin 2t), \ z = e^t (-C_1 - 3C_2 \cos 2t + 3C_3 \sin 2t).$
742. $x = C_1 e^{2t} + e^{3t} (C_2 \cos t + C_3 \sin t), \quad y = e^{3t} [(C_2 + C_3)\cos t +$
$+ (C_3 - C_2)\sin t], \ z = C_1 e^{2t} + e^{3t} [(2C_2 - C_3)\cos t + (2C_3 + C_2)\sin t].$
743. $x = C_2 \cos t + (C_2 + 2C_3)\sin t, \ y = 2C_1 e^t + C_2 \cos t + (C_2 + 2C_3)\sin t,$
$z = C_1 e^t + C_3 \cos t - (C_2 + C_3)\sin t.$ **744.** $x = C_1 e^{2t} + C_3 e^{3t},$
$y = C_1 e^{2t} + C_2 e^{3t}, \ z = C_1 e^{2t} + C_3 e^{3t}.$ **745.** $x = C_1 + C_2 e^t, \ y = 3C_1 + C_3 e^t,$
$z = -C_1 + (C_2 - C_3) e^t.$ **746.** $x = C_1 e^{3t} + C_2 e^{-t}, \quad y = -C_1 e^{3t} +$
$+ (C_2 + 2C_3) e^{-t}, \ z = -3C_1 e^{3t} + C_3 e^{-t}.$ **747.** $x = C_1 e^{2t} + C_3 e^{-5t},$
$y = C_2 e^{2t} + 3C_3 e^{-5t}, \ z = (C_1 - 2C_2) e^{2t} + 2C_3 e^{-5t}.$ **748.** $x = (C_1 +$
$+ C_2 t) e^t + C_3 e^{2t}, \ y = (C_1 - 2C_2 + C_2 t) e^t, \ z = (C_1 - C_2 + C_2 t) e^t + C_3 e^{2t}.$
749. $x = (C_2 + C_3 t) e^{-t}, \qquad y = 2C_1 e^t - (2C_2 + C_3 + 2C_3 t) e^{-t},$
$z = C_1 e^t - (C_2 + C_3 + C_3 t) e^{-t}.$ **750.** $x = C_1 + C_2 t + 4C_3 e^{3t},$
$y = C_2 - 2C_1 - 2C_2 t + 4C_3 e^{3t}, \qquad z = C_1 - C_2 + C_2 t + C_3 e^{3t}.$
751. $x = (C_1 + C_3 t) e^t, \ y = (C_2 + 2C_3 t) e^t, \ z = (C_1 - C_2 - C_3 - C_3 t) e^t.$
752. $x = (C_1 + C_2 t + C_3 t^2) e^{2t}, \ y = [2C_1 - C_2 + (2C_2 - 2C_3) t + 2C_3 t^2] e^{2t},$
$z = [C_1 - C_2 + 2C_3 + (C_2 - 2C_3) t + C_3 t^2] e^{2t}.$ **753.** $x = 3C_1 e^t +$
$+ 3C_2 e^{-t} + C_3 \cos t + C_4 \sin t, \ y = C_1 e^t + C_2 e^{-t} + C_3 \cos t + C_4 \sin t.$
754. $x = -2e^t (C_1 + C_2 + C_2 t) - 2e^{-t} (C_3 - C_4 + C_4 t), \quad y = e^t (C_1 +$
$+ C_2 t) + e^{-t} (C_3 + C_4 t).$ **755.** $x = e^t (C_1 \cos t + C_2 \sin t) +$
$+ e^{-t} (C_3 \cos t + C_4 \sin t), \qquad y = e^t (C_1 \sin t - C_2 \cos t) +$
$+ e^{-t} (C_4 \cos t - C_3 \sin t).$

756. $x = C_1 e^t + C_2 e^{-t} + C_3 e^{2t} + C_5 e^{-2t}$,
$y = C_1 e^t + C_3 e^{-t} + C_4 e^{2t} + C_6 e^{-2t}$, $z = C_1 e^t + C_2 e^{-t} - (C_3 + C_4) e^{2t} - (C_5 + C_6) e^{-2t}$. **757.** $x = 3C_1 e^t + C_2 e^{-t}$, $y = C_1 e^t + C_2 e^{-t}$.
758. $x = C_1 e^t + C_2 e^{-t} + 2C_3 e^{-2t}$, $y = 2C_1 e^t + C_3 e^{-2t}$. **759.** $x = 3C e^t$, $y = C e^{-t}$. **760.** $x = -2C_2 e^{3t} + C_3 e^t$, $y = C_1 e^{-t} + C_2 e^{3t}$.
761. $x = 2C_1 e^{2t} + 2C_2 e^{-2t} + 2C_3 \cos 2t + 2C_4 \sin 2t$, $y = 3C_1 e^{2t} - 3C_2 e^{-2t} - C_3 \sin 2t + C_4 \cos 2t$. **762.** $x = C_1 e^{\frac{t}{2}} - 4C_2 e^{-2t}$,
$y = C_1 e^{\frac{t}{2}} + C_2 e^{-2t}$. **763.** $x = (C_1 + C_2 t) e^t + C_3 e^{-t}$, $y = (-2C_1 - C_2 - 2C_2 t) e^t - 4C_3 e^{-t}$. **764.** $x = C_1 e^t + C_2 e^{-t} + C_3 e^{2t} + C_4 e^{-2t}$,
$y = C_1 e^t + 5C_2 e^{-t} + 2C_3 e^{2t} + 2C_4 e^{-2t}$. **765.** $x = C_1 + C_2 e^t + C_3 \cos t + C_4 \sin t$,
$y = -C_1 - C_2 e^t + \left(\frac{3}{5} C_4 - \frac{4}{5} C_3\right) \cos t - \left(\frac{3}{5} C_3 + \frac{4}{5} C_4\right) \sin t$.
766. $x = C_1 e^t + C_2 e^{-t} + t e^t - t^2 - 2$, $y = C_1 e^t - C_2 e^{-t} + (t-1) e^t - 2t$.
767. $x = C_1 e^{2t} + C_2 e^{-t} - 2 \sin t - \cos t$, $y = 2C_1 e^{2t} - C_2 e^{-t} + \sin t + 3 \cos t$.
768. $x = C_1 e^t + 2C_2 e^{4t} + 3e^{5t}$, $y = -C_1 e^t + C_2 e^{4t} + e^{5t}$.
769. $x = C_1 e^{-t} + 4C_2 e^{2t} + 3e^{-2t}$, $y = C_1 e^{-t} + C_2 e^{2t} + 4e^{-2t}$.
770. $x = C_1 e^{2t} + C_2 e^{3t} + (t+1) e^{2t}$, $y = -2C_1 e^{2t} - C_2 e^{3t} - 2t e^{2t}$.
771. $x = (C_1 + 2C_2 t) e^t - 3$, $y = (C_1 + C_2 + 2C_2 t) e^t - 2$.
772. $x = C_1 e^{2t} + 3C_2 e^{4t} - e^{-t} - 4e^{3t}$, $y = C_1 e^{2t} + C_2 e^{4t} - 2e^{-t} - 2e^{3t}$.
773. $x = C_1 e^{2t} + C_2 e^{-2t} - \frac{1}{4} - \frac{2}{3} e^t$, $y = C_1 e^{2t} - 3C_2 e^{-2t} - \frac{3}{4} - e^t$.
774. $x = C_1 e^t + 2C_2 e^{2t} - \cos t + 3 \sin t$, $y = -C_1 e^{-t} + C_2 e^{2t} + 2 \cos t - \sin t$.
775. $x = 4C_1 e^t + C_2 e^{-2t} - 4t e^t$, $y = C_1 e^t + C_2 e^{-2t} - (t-1) e^t$.
776. $x = C_1 e^{3t} + 3t^2 + 2t + C_2$, $y = -C_1 e^{3t} + 6t^2 - 2t + 2C_2 - 2$.
777. $x = 2C_1 e^{2t} + C_2 e^{-3t} - (12t + 13) e^t$, $y = C_1 e^{2t} - 2C_2 e^{-3t} - (8t + 6) e^t$.
778. $x = 2C_1 e^{8t} - 2C_2 - 6t + 1$, $y = 3C_1 e^{8t} + C_2 + 3t$.
779. $x = 3C_1 e^t + C_2 e^{-t} + 3 \sin t$, $y = C_1 e^t + C_2 e^{-t} - \cos t + 2 \sin t$.
780. $x = C_1 e^{-t} + C_2 e^{5t} - 3e^t + 2t - \frac{13}{5}$, $y = -C_1 e^{-t} + C_2 e^{5t} + e^t - 3t + \frac{12}{5}$. **781.** $x = (C_1 + C_2 t - t^2) e^t$, $y = [C_1 - C_2 + (C_2 + 2) t - t^2] e^t$.
782. $x = C_1 e^t + 3C_2 e^{2t} + \cos t - 2 \sin t$, $y = C_1 e^t + 2C_2 e^{2t} + 2 \cos t - 2 \sin t$. **783.** $x = C_1 e^t + C_2 e^{3t} + t e^t - e^{4t}$,
$y = -C_1 e^t + C_2 e^{3t} - (t+1) e^t - 2e^{4t}$. **784.** $x = C_1 \cos 2t - C_2 \sin 2t + 2t + 2$,
$y = (C_1 + 2C_2) \cos 2t + (2C_1 - C_2) \sin 2t + 10t$. **785.** $x = C_1 e^t + C_2 e^{3t} + e^t (2 \cos t - \sin t)$, $y = C_1 e^t - C_2 e^{3t} + e^t (3 \cos t + \sin t)$.
786. $x = C_1 \cos t + C_2 \sin t + \operatorname{tg} t$, $y = -C_1 \sin t + C_2 \cos t + 2$.
787. $x = C_1 e^t + 2C_2 e^{2t} - e^t \ln(e^{2t} + 1) + 2e^{2t} \operatorname{arctg} e^t$, $y = C_1 e^t + 3C_2 e^{2t} - e^t \ln(e^{2t} + 1) + 3e^{2t} \operatorname{arctg} e^t$. **788.** $x = C_1 + 2C_2 e^{-t} + 2e^{-t} \ln|e^t - 1|$, $y = -2C_1 - 3C_2 e^{-t} - 3e^{-t} \ln|e^t - 1|$.
789. $x = C_1 \cos t + C_2 \sin t + t (\cos t + \sin t) + (\cos t - \sin t) \ln|\cos t|$, $y = (C_1 - C_2) \cos t + (C_1 + C_2) \sin t + 2 \cos t \ln|\cos t| + 2t \sin t$.
790. $x = \left(C_1 + 2C_2 t - 8t^{\frac{5}{2}}\right) e^t$, $y = \left(C_1 + 2C_2 t - C_2 - 8t^{\frac{5}{2}} + 10t^{\frac{3}{2}}\right) e^t$.

15. Lyapunov Stability.

791. Stable. 792. Stable. 793. Stable. 794. Unstable.

795. Unstable. 796. Stable. 797. The solution (0,0) is un-

stable; the solution (1,2) is stable. 798. The solutions

(1,2), (2,1) are both unstable. 799. The solution (2,1) is

unstable; the solution (2,-1) is stable. 800. If k is an

integer, the solutions $x = 2k\pi$, $y = 0$ are unstable, but the

solutions $x = (2k + 1)\pi$, $y = 0$ are stable. 801. Stable.

802. Stable. 803. No, to both questions. 806. Unstable.

807. Stable. 808. Stable. 809. Stable. 810 Stable.

16. Singular Points.

811. Saddle. 812. Node. 813. Focus. 814. Node.

815. Saddle. 816. Center. 817. Degenerate node. 818. Node.

819. Singular node. 820. Focus. 821. Node. 822. Degenerate

node. 823. Focus. 824. Saddle. 825. Center. 826. Degenerate

node. 827, 828. The singular points fill out a straight line.

829. The point (-2,-1) is a node. 830. The point (1,-2) is a

focus. 831. The point (3,6) is a degenerate node. 832. The

point (2,1) is a saddle; the point (-2,1) is a node. 835. The

point (4,2) is a node; the point (-2,-1) is a focus. 836. The

144

point (1,0) is a saddle; the point (0,2) is a degenerate node.
837. The point (1,0) is a singular node; the point (-1,0) is
a saddle. 838. The point (0,1) is a center; the point (0,-1)
is a saddle. 839. The point (2,2) is a node; the point (0,-2)
is a saddle; (-1-1), a focus. 840. The points (1,0) and (-1,0)
are saddles; the points (0,1) and (0,-1) are centers. 841. Singular points as follows: (1,1) is a saddle; (1,-1) is a node;
(2,2), (-2,2) are foci. 842. (0,0) is a focus; (0,8) is a saddle; (3,-1) is a saddle; (7,1) is a node. 843. In the upper
half-plane y > 0, the integral curves are arranged somewhat
like those around a saddle point; in the lower half-plane
y < 0, like a node. 844. A single curve has a simple cusp at
the origin (0,0). The other curves do not touch the singular
point. 845. In the upper half-plane y > 0, all the integral
curves terminate in the singular point; in the lower half-plane
y < 0, none does. 846. Two integral curves pass through the
singular point, where they are mutually tangent. The remaining
curves form a saddle-point configuration. 847. In the upper
half-plane y > 0, the curves do not enter the singular point.
In the third quadrant y < 0, x < 0, the curves behave like

those near a degenerate node. In the fourth quadrant

$y < 0$, $x > 0$, the curves form a saddle-point configuration

17. Problems in the Theory of Oscillations.

851. There are two cases to consider. If $n^2 > 4km$, the

solution is $x = \dfrac{v_0}{2\gamma}(e^{(-\alpha+\gamma)t} - e^{(-\alpha-\gamma)t})$, $\alpha = \dfrac{n}{2m}$, $\gamma = \dfrac{\sqrt{n^2 - 4km}}{2m}$. If

$n^2 < 4km$, the solution is $x = \dfrac{v_0}{\beta}e^{-\alpha t}\sin\beta t$, $\alpha = \dfrac{n}{2m}$, $\beta = \dfrac{\sqrt{4km - n^2}}{2m}$.

852. $n = \sqrt{4km}$.　860. $\dfrac{1}{2\pi}\sqrt{K\left(\dfrac{1}{I_1} + \dfrac{1}{I_2}\right)}$.

861. $A = \dfrac{B}{1 - \dfrac{m}{k}\omega^2}$.　862. $I = \dfrac{V}{R}\left(1 - e^{-\frac{R}{L}t}\right)$.　863. $I = \dfrac{V}{R}e^{-\frac{t}{RC}}$.

864. $I = \dfrac{q}{RC}e^{-\frac{t}{RC}}$.　　865. $I = \dfrac{q}{\omega CL}e^{-\frac{Rt}{2L}}\sin\omega t$, $CR^2 < 4L$,

$\omega = \dfrac{\sqrt{4CL - R^2C^2}}{2LC}$.　　　　866. $I = A\sin(\omega t - \varphi)$,

$A = \dfrac{V}{\sqrt{R^2 + \omega^2 L^2}}$, $\varphi = \operatorname{arctg}\dfrac{\omega L}{R}$.　　867. $I = A\sin(\omega t - \varphi)$,

$A = \dfrac{V}{\sqrt{R^2 + \left(\omega L - \dfrac{1}{\omega C}\right)^2}}$, $\varphi = \operatorname{arctg}\dfrac{\omega L - \dfrac{1}{\omega C}}{R}$;　$\max A = \dfrac{V}{R}$

where $\omega^2 = \dfrac{1}{LC}$. 868. $I = A\sin(\omega t - \varphi)$. $A = \dfrac{V}{\sqrt{R^2 + \left(\dfrac{\omega L}{1 - \omega^2 LC}\right)^2}}$;

the maximum value of A is V/R; this value occurs for $\omega = 0$

and $\omega = \infty$; the minimum value of A is 0; this value occurs

for $\omega^2 = (LC)^{-1}$. 870. $x(\pi) = -4$. The amplitude decreases by

1 in every half oscillation. 871. $l\ddot{\varphi} + g\sin\varphi = 0$.

872. $ml\ddot{\varphi} + kl^2\dot{\varphi}|\dot{\varphi}| + mg\sin\varphi = 0$.　　873. $\dfrac{d^2 I_L}{dt^2} + \dfrac{R}{L}\dfrac{dI_L}{dt} - \dfrac{1}{CL}f\left(M\dfrac{dI_L}{dt}\right) + \dfrac{I_L}{CL} = 0$:

if RC < M f'(0), the equilibrium $I_L(t)$ = f(0) is unstable.
874. In the latter case, the relation F'(0) < 0 holds, and
there is a limit cycle in the phase plane. 892. a < -1/2;
a > -1/2.

18. Dependence of Solutions on Initial Conditions and on
Parameters.

895. Smaller than 0.03. 896. Smaller than 0.1(-1 + exp 2T).
897. $\left| \tilde{x} - x \right| + \left| \tilde{y} - y \right|$ < 0.0012. 898. The error is less than
0.003. 899. The error is less than 0.034.

900. $e^{2x} - x - 1$. 901. $t^2 \ln t + 2t^2 - 2t$.

902. $\frac{e^{2t}}{36} - \frac{e^{-2t}}{4} + \left(\frac{2}{9} - \frac{t}{3} \right) e^{-t} + \frac{1}{8}$. 905. $x = \sin t +$

$+ \mu \left(\frac{1}{6} - \frac{1}{2} \cos 2t \right) + \mu^2 \left(\frac{1}{2} \sin t - \frac{1}{6} \sin 3t \right) + O(\mu^3)$. 906. $x = \cos 2t +$

$+ \mu \left(\frac{1}{10} - \frac{1}{22} \cos 4t \right) + \mu^2 \left(\frac{17}{110} \cos 2t + \frac{1}{682} \cos 6t \right) + O(\mu^3)$.

907. $x = \mu \cos t + \mu^3 \left(-\frac{3}{8} \cos t + \frac{1}{24} \cos 3t \right) + O(\mu^5)$. 908. $x_1 = 1 +$

$+ \mu \sin t - \frac{\mu^2}{4} (1 + \cos 2t) + O(\mu^3)$, $x_2 = -1 - \frac{\mu}{3} \sin t +$

$+ \frac{\mu^2}{36} \left(1 - \frac{1}{3} \cos 2t \right) + O(\mu^3)$. 909. $x_1 = -\frac{\mu}{5} \sin 2t + \frac{\mu^3}{648} \left(\sin 2t - \right.$

$\left. - \frac{1}{35} \sin 6t \right) + O(\mu^5)$, $x_2 = \pi - \frac{\mu}{5} \sin 2t - \frac{\mu^3}{1000} \left(\frac{1}{5} \sin 2t - \frac{1}{111} \sin 6t \right) +$

$+ O(\mu^5)$. 910. $x = \frac{1}{8} \sin t + \frac{1}{3} \sin 2t - \frac{1}{8} \sin 3t + O(\mu)$.

911. $x = 2\mu^{\frac{1}{3}} \sin t - \mu \left(\frac{1}{12} \sin t + \frac{1}{4} \sin 3t \right) + O\left(\mu^{\frac{5}{3}} \right)$. 912. $x = C \cos \tau +$

$+ C^2 \left(\frac{1}{2} - \frac{1}{3} \cos \tau - \frac{1}{6} \cos 2\tau \right) + O(C^3)$, $\tau = t \left(1 - \frac{5}{12} C^2 + O(C^3) \right) + C_2$.

913. $x = C \cos \tau + \frac{C^3}{192} (\cos \tau - \cos 3\tau) + O(C^5)$, $\tau = t \left(1 - \frac{C^2}{16} + \right.$

$\left. + O(C^4) \right) + C_2$. 914. $x = 2 \cos \tau - \frac{\mu}{4} \sin 3\tau + O(\mu^2)$,

$\tau = t \left(1 - \frac{\mu^2}{16} + O(\mu^4) \right) + C$.

19. Non-Linear Systems.

921. $y = C_2 e^{C_1 x^2}$, $\quad z = \dfrac{1}{2C_1 C_2}\, e^{-C_1 x^2}$. $\qquad$ **922.** $y = C_2 e^{C_1 x}$,

$z = x + \dfrac{C_2}{C_1}\, e^{C_1 x}$. $\qquad$ **923.** $y = \dfrac{x + C_1}{x + C_2}$, $\quad z = \dfrac{(C_2 - C_1)\, x}{(x + C_2)^2}$.

924. $y = C_2 e^{C_1 x^2}$, $z = \dfrac{2C_1}{C_2}\, x e^{-C_1 x^2}$. $\quad$ **925.** $y = -\dfrac{1}{C_1} + \dfrac{C_1}{2}\, (x + C_2) -$

$-\dfrac{C_1}{4}\, (x + C_2)^2$, $z = \dfrac{C_1}{4}\, (x + C_2)^2 + \dfrac{1}{C_1}$. **926.** $y = C_1 z$, $x = 2y - z + C_2$.

927. $x^2 - y^2 = C_1$, $\quad x + y = C_2 z$. $\qquad$ **928.** $x - y = C_1\, (y - z)$,

$(x + y + z)\, (x - y)^2 = C_2$. **929.** $x + z = C_1$, $(x + y + z)\, (y - 3x - z) = C_2$.

930. $x^2 - z^2 = C_1$, $y^2 - u^2 = C_2$, $(x + z) = C_3\, (u + y)$. **931.** $x + z = C_1$,

$y + u = C_2$, $\quad (x - z)^2 + (y - u)^2 = C_3$. $\qquad$ **932.** $x^2 - 2y = C_1$,

$6xy - 2x^3 - 3z^2 = C_2$. **933.** $y^2 + z^2 = C_1$, $x - yz = C_2$. **934.** $x = C_1 y$,

$xy - z = C_2 x$. **935** $x = C_1 y$, $\quad xy - 2\sqrt{z^2 + 1} = C_2$. $\quad$ **936.** $y = C_1 z$,

$x - y^2 - z^2 = C_2 z$. **937.** $y^2 + z^2 = C_1$, $x\, (y - z) = C_2$. **938.** $xz = C_1$,

$xy + z^2 = C_2$. **939.** $x + z - y = C_1$, $\ln|x| + \dfrac{z}{y} = C_2$. **940.** $x^2 + y^2 +$

$+ z^2 = C_1$, $yz = C_2 x$. $\qquad$ **941.** 1: yes; 2: no. **942.** 1: no; 2: yes.

943. Yes. **944.** Dependent.

20. First Order Partial Differential Equations.

946. $F(x^2 - y^2, x - y + z) = 0$. **947.** $F\left(e^{-x} - y^{-1}, z + \dfrac{x - \ln|y|}{e^{-x} - y^{-1}}\right) = 0$.

948. $F\left(x^2 - 4z, \dfrac{(x + y)^2}{x}\right) = 0$. $\qquad$ **949.** $F\left(x^2 + y^2, \dfrac{z}{x}\right) = 0$.

950. $F\left(\dfrac{x^2}{y}, xy - \dfrac{3z}{x}\right) = 0$. **951.** $F\left(\dfrac{1}{x + y} + \dfrac{1}{z}, \dfrac{1}{x - y} + \dfrac{1}{z}\right) = 0$.

952. $F(x^2 + y^4, y(z + \sqrt{z^2 + 1})) = 0$. **953.** $F\left(\dfrac{1}{x} - \dfrac{1}{y}, \ln|xy| - \dfrac{z^2}{2}\right) = 0$.

954. $F\left(x^2 + y^2, \operatorname{arctg} \dfrac{x}{y} + (z + 1)\, e^{-z}\right) = 0$. $\qquad$ **955.** $F(z^2 - y^2,$

$x^2 + (y - z)^2) = 0$. **956.** $F\left(\dfrac{z}{x}, 2x - 4z - y^2\right) = 0$. **957.** $F(z - \ln|x|,$

$2x\, (z - 1) - y^2) = 0$. **958.** $F(\operatorname{tg} z + \operatorname{ctg} x, 2y + 2\operatorname{tg} z \cdot \operatorname{ctg} x + \operatorname{ctg}^2 x) = 0$.

959. $F\left(\dfrac{x + y + z}{(x - y)^2}, (x - y)(x + y - 2z)\right) = 0$. **960.** $F((x - y)(z + 1),$

$(x + y)\, (z - 1)) = 0$. $\quad$ **961.** $F\left(u\, (x - y), u\, (y - z), \dfrac{x + y + z}{u^2}\right) =$

962. $F\left(\dfrac{x}{y}, xy - 2u, \dfrac{z + u - xy}{x}\right) = 0$. **963.** $F\left(\dfrac{x - y}{z}, (2u + x + y)\, z, \right.$

$\left. \dfrac{u - x - y}{z^2}\right) = 0$. $\qquad$ **964.** $y^2 - x^2 - \ln\sqrt{y^2 - x^2} = z - \ln|y|$.

965. $2x^2\, (y + 1) = y^2 + 4z - 1$. $\qquad$ **966.** $(x + 2y)^2 = 2x\, (z + xy)$.

967. $\sqrt{\dfrac{z}{y^3}}\ \sin x = \sin \sqrt{\dfrac{z}{y}}\,.$ **968.** $2xy + 1 = x + 3y + z^{-1}.$

969. $x - 2y = x^2 + y^2 + z.$ **970.** $2x^2 - y^2 - z^2 = a^2.$

971. $[(y^2z - 2)^2 - x^2 + z]\, y^2 z = 1.$ **972.** $x^2 + z^2 = 5\,(xz - y).$

973. $3\,(x + y + z)^2 = x^2 + y^2 + z^2.$ **974.** $xz = (xz - y - x + 2z)^2.$

975. $(1 + yz)^3 = 3yz\,(1 + yz - x) + y^3.$ **976.** $x + y + z = 0.$

977. $2\,(x^3 - 4z^3 - 3yz)^2 = 9\,(y + z^2)^3.$ **978.** $(x - y)\,(3x + y + 4z) = 4z.$

979. $xz + y^2 = 0.$ **980.** $z = xy + f(y/x)$, where f is any

differentiable function satisfying $f(1) = 0.$

981. $F(x^2 - y^2,\ 2x^2 + z^2) = 0.$ **981.** $F(x^2 - y^2,\ 2x^2 + z^2) = 0.$

983. $F(2x - z,\ x - y) = 0.$ **984.** $(x - y)^2 + (z - x - y)^2 = 4.$

985. $F\!\left(\dfrac{y - b}{x - a},\ \dfrac{z - c}{x - a}\right) = 0.$ **986.** $F\!\left(\dfrac{x^2}{y},\ \dfrac{z}{y}\right) = 0.$ **987.** $z = Cxy^2.$

988. No solution.   **989.** $z = 0.$ **990.** No solution.   **991.** $x^3 y^2 z = C.$

992. $z = y^2 - xy.$ **993.** $x^2 yz = C - x^3;\ x = 0.$

MISCELLANEOUS FORMULAS

$\sin (A \pm B) = \sin A \cos B \pm \cos A \sin B$

$\cos (A \pm B) = \cos A \cos B \mp \sin A \sin B$

$\tan (A \pm B) = \dfrac{\tan A \pm \tan B}{1 \mp \tan A \tan B}$

$\sin 3A = 3 \sin A - 4 \sin^3 A$

$\cos 3A = 4 \cos^3 A - 3 \cos A$

$\sin {}^P A = 2^{-p} \displaystyle\sum_{t=o}^{p} \binom{p}{t} (-1)^t \sin (p-2t)A$

$\cos {}^P A = 2^{-p} \displaystyle\sum_{t=o}^{p} \binom{p}{t} \cos (p-2t)A,$

$\binom{p}{t} = \dfrac{p!}{t! \ (p-t)!} \ , \ 0! = 1.$

$\sin A + \sin B = 2 \sin \dfrac{1}{2} (A + B) \cos \dfrac{1}{2} (A - B)$

$\sin D \cos E = \dfrac{1}{2} \sin (D + E) + \dfrac{1}{2} \sin (D - E)$

$\sin D \sin E = \dfrac{1}{2} \cos (D - E) - \dfrac{1}{2} \cos (D + E)$

$\sinh (A + B) = \sinh A \cosh B + \cosh A \sinh B$

$\cosh (A + B) = \cosh A \cosh B + \sinh A \sinh B$

If $A = \cosh B$, than $B = LN \ (A + \sqrt{A^2 - 1})$

If $A = \sinh B$, than $B = LN \ (A + \sqrt{A^2 + 1})$

FORMULAS OF DIFFERENTIATION

Function	Derivative (evaluated when independent variable has the value x)
$f(x)$	$\underset{h \to 0,\ h \neq 0}{\text{LIM}} \dfrac{f(x + h) - f(x)}{h}$, if this limit exists
Constant	0
x^p	$p\, x^{p-1}$
$\sqrt{x}$	$\dfrac{1}{2\sqrt{x}}$
$\sin x$	$\cos x$
$\cos x$	$-\sin x$
$\tan x = tg\ x$	$(\sec x)(\sec x)$
$\cotan x = ctg\ x$	$-(\csc x)(\csc x)$
$\sec x$	$(\sec x)(\tan x)$
$\cosec x$	$-(\cosec x)(\cotan x)$
$\exp x = e^x$	$\exp x$
$LOG\ x = LN\ x$	$1/x = x^{-1}$
$LOG_a x$	$\dfrac{1}{x\ LOG\ a} = \dfrac{1}{x\ LN\ a}$
$\sinh x$	$\cosh x$

FORMULAS OF DIFFERENTIATION (Continued)

Function	Derivative (evaluated when independent variable has the value x)
$\cosh x$	$\sinh x$
$\tanh x$	$(\text{sech } x)(\text{sech } x)$
$\text{Arcsin } x$	$\dfrac{1}{\sqrt{1 - x^2}}$
$\text{Arccos } x$	$\dfrac{-1}{\sqrt{1 - x^2}}$
$\text{Arctan } x$	$\dfrac{1}{1 + x^2}$
$\text{Arctanh } x$	$\dfrac{1}{1 - x^2}$
$\text{Arcsinh } x$	$\dfrac{1}{\sqrt{1 + x^2}}$
$\text{Arccosh } x$	$\dfrac{1}{\sqrt{x^2 - 1}}$
$u\,v$	$u\,v' + u'\,v$
$u/v = u\,v^{-1}$	$(u'v - u\,v')\,v^{-2}$
$u^m v^n$	$(n\,u\,v' + m\,u'\,v)(u^{m-1}v^{n-1})$

$$D_x u = (D_t u)(D_x t)$$

SOME COMMON INTEGRALS

Function of u	Integral of the function with respect to u: $\int f(u)\,du$
0	C = arbitrary constant
u^p (p≠-1)	$u^{p+1}/(p+1)$ + C
u^{-1}	LN $\lvert u \rvert$ + C
sin u	- cos u + C
cos u	sin u + C
tan u = tg u	LN $\lvert$ sec u $\rvert$ + C
cotan u = ctg u	LN $\lvert$ sin u $\rvert$ + C
sec u	LN $\lvert$ sec u + tan u $\rvert$ + C
csc u	LN $\lvert$ csc u - ctn u $\rvert$ + C
exp u = e^u	exp u + C
LN u	u LN u - u + C
sinh u	cosh u + C
cosh u	sinh u + C
tanh u = sinh u/cosh u	LN cosh u + C
cotanh u = cosh u/sinh u	LN $\lvert$ sinh u $\rvert$ + C
$\cos^2 u$	$\frac{1}{2} u + \frac{1}{4} \sin 2u$ + C

SOME COMMON INTEGRALS (Continued)

Function of u

Integral of the function with respect to u: $\int f(u)\,du$

$\sin^2 u = 1 - \cos^2 u$

$\dfrac{1}{2} u - \dfrac{1}{4} \sin 2u + C$

$\sin u \cos u = \dfrac{1}{2} \sin 2u$

$-\dfrac{1}{4} \cos 2u \qquad + C$

$\dfrac{1}{a + bu}$

$LN \left| a + bu \right|^{1/b} + C$

$= \dfrac{1}{ab} \operatorname{Arc\,tanh} \dfrac{bu}{a} + C$

$\dfrac{1}{\sqrt{a^2 - b^2 u^2}} \qquad (a > 0)$

$\dfrac{1}{b} \operatorname{Arc\,sin} \dfrac{bu}{a} + C$

$\dfrac{1}{a^2 + b^2 u^2}$

$\dfrac{1}{ab} \operatorname{Arc\,tan} \dfrac{bu}{a} + C$

$\dfrac{1}{\sqrt{a^2 + b^2 u^2}} \qquad (a > 0)$

$\dfrac{1}{b} \operatorname{Arc\,sinh} \dfrac{bu}{a} + C$

$= \dfrac{1}{b} LN \left| \sqrt{a^2 + b^2 u^2} + bu \right| + C$

$= -\dfrac{1}{b} LN \left| \sqrt{a^2 + b^2 u^2} - bu \right| + C_1$

$\dfrac{1}{\sqrt{b^2 u^2 - a^2}} \qquad (a > 0)$

$\dfrac{1}{b} \operatorname{Arc\,cosh} \dfrac{bu}{a} + C$

$= \dfrac{1}{b} LN \left| \sqrt{b^2 u^2 - a^2} + bu \right| + C$

SQUARE ROOTS - CUBE ROOTS

N	$\sqrt{N}$	$\sqrt{10N}$	$\sqrt[3]{N}$	$\sqrt[3]{10N}$	$\sqrt[3]{100N}$
0.1	0.316	1.	.464	1.	2.154
0.15	.387	1.225	.531	1.145	2.467
0.2	.447	1.414	.585	1.260	2.714
0.25	.5	1.581	.630	1.357	2.924
0.3	.548	1.732	.669	1.442	3.107
0.4	.632	2.	.739	1.587	3.420
0.5	.707	2.236	.794	1.710	3.684
0.6	.775	2.449	.843	1.817	3.915
0.7	.837	2.646	.888	1.913	4.121
0.8	.894	2.828	.928	2.	4.309
0.9	0.949	3.	.965	2.080	4.481
1.0	1.	3.162	1.	2.154	4.641

$\pi = 3.14159$ $e = 2.71828$

$\pi^2 = 9.870$

LOGARITHMS

N	$\text{LOG}_e N = \text{LN } N$	$\text{LOG}_{10} N$
0.01	$-\ 4.605$	
0.05	$-\ 2.996$	
0.1	$-\ 2.303$	
0.5	$-\ 0.693$	
1	$+\ 0$	0
1.5	$+0.40547$	.1761
2	$+0.69315$	.3010
2.5	$+0.91629$	.3979
e	$+1.$	0.43429
3	$+1.09861$	.4771
π	$+1.14473$	.4971
4	$+1.38629$	.6021
5	$+1.60944$	.6990
6	$+1.79176$	.7781
7	$+1.94951$	.8451
8	$+2.08944$	.9031
9	$+2.19722$	.9542
10	$+2.30259$	1.
100	$+4.60517$	
1000	$+6.9077$	

$$\text{LOG}_e N = (2.30259)\ \text{LOG}_{10} N$$

EXPONENTIALS

$$\text{EXP } N = 1 + N = \frac{N^2}{2} + \frac{N^3}{6} + \dots$$

$$\text{EXP } -N = 1 - N + \frac{N^2}{2} - \frac{N^3}{6} + \dots$$

0.01	1.0101	.9901
2	1.0202	.9802
3	1.0305	.9704
4	1.0408	.9608
5	1.0513	.9512
6	1.0618	.9418
7	1.0725	.9324
8	1.0833	.9231
0.09	1.0942	.9139
0.1	1.1052	.9048
2	1.2214	.8187
3	1.3499	.7408
4	1.4918	.6703
5	1.6487	.6065
6	1.8221	.5488
7	2.0138	.4966
8	2.2255	.4493
0.9	2.4596	.4066
1.	2.7183	.3679
2.	7.3891	.1353
3.	20.086	.0498
4.	54.598	.0183
5.	148.41	.0067
6.	403.43	.0025
7.	1096.6	.0009
8.	2981.0	.0003
9.	8103.	.00012
10.	22026.	.00004

TRIGONOMETRIC FUNCTIONS

RADIAN VALUES OF THE ARGUMENT

t	SIN t	COS t	TAN t	CTN t	SEC t	CSC t
0.1	+.0998	+.9950	+ .1003	9.97	1.005	10.02
.2	.1987	.9801	.2027	4.93	1.020	5.033
.3	.2955	.9553	.3093	3.23	1.047	3.384
.4	.3894	.9211	.4228	2.37	1.086	2.568
.5	.4794	.8776	.5463	1.83	1.139	2.086
.6	.5646	.8253	.6841	1.462	1.212	1.771
.7	.6442	.7648	.8423	1.187	1.307	1.552
.8	.7174	.6967	1.030	.9712	1.435	1.394
0.9	.7833	.6216	1.260	.7936	1.609	1.277
1.0	.8415	.5403	1.557	.6421	1.851	1.188
1.1	.8912	.4536	1.965	.5090	2.205	1.122
1.2	.9320	.3624	2.57	.3888	2.760	1.073
1.3	.9636	.2675	3.60	.2776	3.738	1.038
1.4	.9854	.1700	5.80	.1725	5.883	1.015
1.5	.9975	+.0707	14.1	.0709	14.14	1.003
1.6	+.9996	-.0292	-34.	-.0292	-34.25	1.000
1.7	.9917	-.1288	- 7.70	-.1299	- 7.761	1.008
1.8	.9738	-.2272	- 4.29	-.2333	- 4.401	1.027
1.9	.9463	-.3233	- 2.93	-.3416	- 3.093	1.057
2.0	.9093	-.4161	- 2.19	-.4577	- 2.403	1.100
2.1	.8632	-.5048	- 1.710	-.5848	- 1.981	1.158
2.2	.8085	-.5885	- 1.374	-.7279	- 1.699	1.237
2.5	+.5985	-.8011	- .7470	-1.339	- 1.248	1.671
3.0	+.1411	-.9900	- .1425	-7.02	- 1.010	7.086
3.5	-.3508	-.9365	.3746	2.67	- 1.068	-2.851
4.0	-.7568	-.6536	1.158	.8637	- 1.530	-1.321
4.5	-.9775	-.2108	4.64	.2156	- 4.744	-1.023
5.0	-.9589	+.2837	-3.38	- .2958	3.525	-1.043

1 RAD = 180/π DEG = 57. 29578 DEG = 206 264.81 SECONDS OF ARC

A CATALOG OF SELECTED
DOVER BOOKS
IN SCIENCE AND MATHEMATICS

Mathematics–Bestsellers

HANDBOOK OF MATHEMATICAL FUNCTIONS: with Formulas, Graphs, and Mathematical Tables, Edited by Milton Abramowitz and Irene A. Stegun. A classic resource for working with special functions, standard trig, and exponential logarithmic definitions and extensions, it features 29 sets of tables, some to as high as 20 places. 1046pp. 8 x 10 1/2. 0-486-61272-4

ABSTRACT AND CONCRETE CATEGORIES: The Joy of Cats, Jiri Adamek, Horst Herrlich, and George E. Strecker. This up-to-date introductory treatment employs category theory to explore the theory of structures. Its unique approach stresses concrete categories and presents a systematic view of factorization structures. Numerous examples. 1990 edition, updated 2004. 528pp. 6 1/8 x 9 1/4. 0-486-46934-4

MATHEMATICS: Its Content, Methods and Meaning, A. D. Aleksandrov, A. N. Kolmogorov, and M. A. Lavrent'ev. Major survey offers comprehensive, coherent discussions of analytic geometry, algebra, differential equations, calculus of variations, functions of a complex variable, prime numbers, linear and non-Euclidean geometry, topology, functional analysis, more. 1963 edition. 1120pp. 5 3/8 x 8 1/2. 0-486-40916-3

INTRODUCTION TO VECTORS AND TENSORS: Second Edition--Two Volumes Bound as One, Ray M. Bowen and C.-C. Wang. Convenient single-volume compilation of two texts offers both introduction and in-depth survey. Geared toward engineering and science students rather than mathematicians, it focuses on physics and engineering applications. 1976 edition. 560pp. 6 1/2 x 9 1/4. 0-486-46914-X

AN INTRODUCTION TO ORTHOGONAL POLYNOMIALS, Theodore S. Chihara. Concise introduction covers general elementary theory, including the representation theorem and distribution functions, continued fractions and chain sequences, the recurrence formula, special functions, and some specific systems. 1978 edition. 272pp. 5 3/8 x 8 1/2. 0-486-47929-3

ADVANCED MATHEMATICS FOR ENGINEERS AND SCIENTISTS, Paul DuChateau. This primary text and supplemental reference focuses on linear algebra, calculus, and ordinary differential equations. Additional topics include partial differential equations and approximation methods. Includes solved problems. 1992 edition. 400pp. 7 1/2 x 9 1/4. 0-486-47930-7

PARTIAL DIFFERENTIAL EQUATIONS FOR SCIENTISTS AND ENGINEERS, Stanley J. Farlow. Practical text shows how to formulate and solve partial differential equations. Coverage of diffusion-type problems, hyperbolic-type problems, elliptic-type problems, numerical and approximate methods. Solution guide available upon request. 1982 edition. 414pp. 6 1/8 x 9 1/4. 0-486-67620-X

VARIATIONAL PRINCIPLES AND FREE-BOUNDARY PROBLEMS, Avner Friedman. Advanced graduate-level text examines variational methods in partial differential equations and illustrates their applications to free-boundary problems. Features detailed statements of standard theory of elliptic and parabolic operators. 1982 edition. 720pp. 6 1/8 x 9 1/4. 0-486-47853-X

LINEAR ANALYSIS AND REPRESENTATION THEORY, Steven A. Gaal. Unified treatment covers topics from the theory of operators and operator algebras on Hilbert spaces; integration and representation theory for topological groups; and the theory of Lie algebras, Lie groups, and transform groups. 1973 edition. 704pp. 6 1/8 x 9 1/4. 0-486-47851-3

Browse over 9,000 books at www.doverpublications.com

A SURVEY OF INDUSTRIAL MATHEMATICS, Charles R. MacCluer. Students learn how to solve problems they'll encounter in their professional lives with this concise single-volume treatment. It employs MATLAB and other strategies to explore typical industrial problems. 2000 edition. 384pp. 5 3/8 x 8 1/2.   0-486-47702-9

NUMBER SYSTEMS AND THE FOUNDATIONS OF ANALYSIS, Elliott Mendelson. Geared toward undergraduate and beginning graduate students, this study explores natural numbers, integers, rational numbers, real numbers, and complex numbers. Numerous exercises and appendixes supplement the text. 1973 edition. 368pp. 5 3/8 x 8 1/2.   0-486-45792-3

A FIRST LOOK AT NUMERICAL FUNCTIONAL ANALYSIS, W. W. Sawyer. Text by renowned educator shows how problems in numerical analysis lead to concepts of functional analysis. Topics include Banach and Hilbert spaces, contraction mappings, convergence, differentiation and integration, and Euclidean space. 1978 edition. 208pp. 5 3/8 x 8 1/2.   0-486-47882-3

FRACTALS, CHAOS, POWER LAWS: Minutes from an Infinite Paradise, Manfred Schroeder. A fascinating exploration of the connections between chaos theory, physics, biology, and mathematics, this book abounds in award-winning computer graphics, optical illusions, and games that clarify memorable insights into self-similarity. 1992 edition. 448pp. 6 1/8 x 9 1/4.   0-486-47204-3

SET THEORY AND THE CONTINUUM PROBLEM, Raymond M. Smullyan and Melvin Fitting. A lucid, elegant, and complete survey of set theory, this three-part treatment explores axiomatic set theory, the consistency of the continuum hypothesis, and forcing and independence results. 1996 edition. 336pp. 6 x 9.   0-486-47484-4

DYNAMICAL SYSTEMS, Shlomo Sternberg. A pioneer in the field of dynamical systems discusses one-dimensional dynamics, differential equations, random walks, iterated function systems, symbolic dynamics, and Markov chains. Supplementary materials include PowerPoint slides and MATLAB exercises. 2010 edition. 272pp. 6 1/8 x 9 1/4.   0-486-47705-3

ORDINARY DIFFERENTIAL EQUATIONS, Morris Tenenbaum and Harry Pollard. Skillfully organized introductory text examines origin of differential equations, then defines basic terms and outlines general solution of a differential equation. Explores integrating factors; dilution and accretion problems; Laplace Transforms; Newton's Interpolation Formulas, more. 818pp. 5 3/8 x 8 1/2.   0-486-64940-7

MATROID THEORY, D. J. A. Welsh. Text by a noted expert describes standard examples and investigation results, using elementary proofs to develop basic matroid properties before advancing to a more sophisticated treatment. Includes numerous exercises. 1976 edition. 448pp. 5 3/8 x 8 1/2.   0-486-47439-9

THE CONCEPT OF A RIEMANN SURFACE, Hermann Weyl. This classic on the general history of functions combines function theory and geometry, forming the basis of the modern approach to analysis, geometry, and topology. 1955 edition. 208pp. 5 3/8 x 8 1/2.   0-486-47004-0

THE LAPLACE TRANSFORM, David Vernon Widder. This volume focuses on the Laplace and Stieltjes transforms, offering a highly theoretical treatment. Topics include fundamental formulas, the moment problem, monotonic functions, and Tauberian theorems. 1941 edition. 416pp. 5 3/8 x 8 1/2.   0-486-47755-X

Mathematics-Logic and Problem Solving

PERPLEXING PUZZLES AND TANTALIZING TEASERS, Martin Gardner. Ninety-three riddles, mazes, illusions, tricky questions, word and picture puzzles, and other challenges offer hours of entertainment for youngsters. Filled with rib-tickling drawings. Solutions. 224pp. 5 3/8 x 8 1/2.　　　　　　　　　0-486-25637-5

MY BEST MATHEMATICAL AND LOGIC PUZZLES, Martin Gardner. The noted expert selects 70 of his favorite "short" puzzles. Includes The Returning Explorer, The Mutilated Chessboard, Scrambled Box Tops, and dozens more. Complete solutions included. 96pp. 5 3/8 x 8 1/2.　　　　　　　　　　　0-486-28152-3

THE LADY OR THE TIGER?: and Other Logic Puzzles, Raymond M. Smullyan. Created by a renowned puzzle master, these whimsically themed challenges involve paradoxes about probability, time, and change; metapuzzles; and self-referentiality. Nineteen chapters advance in difficulty from relatively simple to highly complex. 1982 edition. 240pp. 5 3/8 x 8 1/2.　　　　　　　　　　0-486-47027-X

SATAN, CANTOR AND INFINITY: Mind-Boggling Puzzles, Raymond M. Smullyan. A renowned mathematician tells stories of knights and knaves in an entertaining look at the logical precepts behind infinity, probability, time, and change. Requires a strong background in mathematics. Complete solutions. 288pp. 5 3/8 x 8 1/2.

0-486-47036-9

THE RED BOOK OF MATHEMATICAL PROBLEMS, Kenneth S. Williams and Kenneth Hardy. Handy compilation of 100 practice problems, hints and solutions indispensable for students preparing for the William Lowell Putnam and other mathematical competitions. Preface to the First Edition. Sources. 1988 edition. 192pp. 5 3/8 x 8 1/2.　　　　　　　　　　　　　　0-486-69415-1

KING ARTHUR IN SEARCH OF HIS DOG AND OTHER CURIOUS PUZZLES, Raymond M. Smullyan. This fanciful, original collection for readers of all ages features arithmetic puzzles, logic problems related to crime detection, and logic and arithmetic puzzles involving King Arthur and his Dogs of the Round Table. 160pp. 5 3/8 x 8 1/2.

0-486-47435-6

UNDECIDABLE THEORIES: Studies in Logic and the Foundation of Mathematics, Alfred Tarski in collaboration with Andrzej Mostowski and Raphael M. Robinson. This well-known book by the famed logician consists of three treatises: "A General Method in Proofs of Undecidability," "Undecidability and Essential Undecidability in Mathematics," and "Undecidability of the Elementary Theory of Groups." 1953 edition. 112pp. 5 3/8 x 8 1/2.　　　　　　　　　　　0-486-47703-7

LOGIC FOR MATHEMATICIANS, J. Barkley Rosser. Examination of essential topics and theorems assumes no background in logic. "Undoubtedly a major addition to the literature of mathematical logic." – *Bulletin of the American Mathematical Society.* 1978 edition. 592pp. 6 1/8 x 9 1/4.　　　　　　　　　　0-486-46898-4

INTRODUCTION TO PROOF IN ABSTRACT MATHEMATICS, Andrew Wohlgemuth. This undergraduate text teaches students what constitutes an acceptable proof, and it develops their ability to do proofs of routine problems as well as those requiring creative insights. 1990 edition. 384pp. 6 1/2 x 9 1/4.　　0-486-47854-8

FIRST COURSE IN MATHEMATICAL LOGIC, Patrick Suppes and Shirley Hill. Rigorous introduction is simple enough in presentation and context for wide range of students. Symbolizing sentences; logical inference; truth and validity; truth tables; terms, predicates, universal quantifiers; universal specification and laws of identity; more. 288pp. 5 3/8 x 8 1/2.　　　　　　　　　　0-486-42259-3

Browse over 9,000 books at www.doverpublications.com

Mathematics–Algebra and Calculus

VECTOR CALCULUS, Peter Baxandall and Hans Liebeck. This introductory text offers a rigorous, comprehensive treatment. Classical theorems of vector calculus are amply illustrated with figures, worked examples, physical applications, and exercises with hints and answers. 1986 edition. 560pp. 5 3/8 x 8 1/2.   0-486-46620-5

ADVANCED CALCULUS: An Introduction to Classical Analysis, Louis Brand. A course in analysis that focuses on the functions of a real variable, this text introduces the basic concepts in their simplest setting and illustrates its teachings with numerous examples, theorems, and proofs. 1955 edition. 592pp. 5 3/8 x 8 1/2.   0-486-44548-8

ADVANCED CALCULUS, Avner Friedman. Intended for students who have already completed a one-year course in elementary calculus, this two-part treatment advances from functions of one variable to those of several variables. Solutions. 1971 edition. 432pp. 5 3/8 x 8 1/2.   0-486-45795-8

METHODS OF MATHEMATICS APPLIED TO CALCULUS, PROBABILITY, AND STATISTICS, Richard W. Hamming. This 4-part treatment begins with algebra and analytic geometry and proceeds to an exploration of the calculus of algebraic functions and transcendental functions and applications. 1985 edition. Includes 310 figures and 18 tables. 880pp. 6 1/2 x 9 1/4.   0-486-43945-3

BASIC ALGEBRA I: Second Edition, Nathan Jacobson. A classic text and standard reference for a generation, this volume covers all undergraduate algebra topics, including groups, rings, modules, Galois theory, polynomials, linear algebra, and associative algebra. 1985 edition. 528pp. 6 1/8 x 9 1/4.   0-486-47189-6

BASIC ALGEBRA II: Second Edition, Nathan Jacobson. This classic text and standard reference comprises all subjects of a first-year graduate-level course, including in-depth coverage of groups and polynomials and extensive use of categories and functors. 1989 edition. 704pp. 6 1/8 x 9 1/4.   0-486-47187-X

CALCULUS: An Intuitive and Physical Approach (Second Edition), Morris Kline. Application-oriented introduction relates the subject as closely as possible to science with explorations of the derivative; differentiation and integration of the powers of x; theorems on differentiation, antidifferentiation; the chain rule; trigonometric functions; more. Examples. 1967 edition. 960pp. 6 1/2 x 9 1/4.   0-486-40453-6

ABSTRACT ALGEBRA AND SOLUTION BY RADICALS, John E. Maxfield and Margaret W. Maxfield. Accessible advanced undergraduate-level text starts with groups, rings, fields, and polynomials and advances to Galois theory, radicals and roots of unity, and solution by radicals. Numerous examples, illustrations, exercises, appendixes. 1971 edition. 224pp. 6 1/8 x 9 1/4.   0-486-47723-1

AN INTRODUCTION TO THE THEORY OF LINEAR SPACES, Georgi E. Shilov. Translated by Richard A. Silverman. Introductory treatment offers a clear exposition of algebra, geometry, and analysis as parts of an integrated whole rather than separate subjects. Numerous examples illustrate many different fields, and problems include hints or answers. 1961 edition. 320pp. 5 3/8 x 8 1/2.   0-486-63070-6

LINEAR ALGEBRA, Georgi E. Shilov. Covers determinants, linear spaces, systems of linear equations, linear functions of a vector argument, coordinate transformations, the canonical form of the matrix of a linear operator, bilinear and quadratic forms, and more. 387pp. 5 3/8 x 8 1/2.   0-486-63518-X

Mathematics–Probability and Statistics

BASIC PROBABILITY THEORY, Robert B. Ash. This text emphasizes the probabilistic way of thinking, rather than measure-theoretic concepts. Geared toward advanced undergraduates and graduate students, it features solutions to some of the problems. 1970 edition. 352pp. 5 3/8 x 8 1/2. 0-486-46628-0

PRINCIPLES OF STATISTICS, M. G. Bulmer. Concise description of classical statistics, from basic dice probabilities to modern regression analysis. Equal stress on theory and applications. Moderate difficulty; only basic calculus required. Includes problems with answers. 252pp. 5 5/8 x 8 1/4. 0-486-63760-3

OUTLINE OF BASIC STATISTICS: Dictionary and Formulas, John E. Freund and Frank J. Williams. Handy guide includes a 70-page outline of essential statistical formulas covering grouped and ungrouped data, finite populations, probability, and more, plus over 1,000 clear, concise definitions of statistical terms. 1966 edition. 208pp. 5 3/8 x 8 1/2. 0-486-47769-X

GOOD THINKING: The Foundations of Probability and Its Applications, Irving J. Good. This in-depth treatment of probability theory by a famous British statistician explores Keynesian principles and surveys such topics as Bayesian rationality, corroboration, hypothesis testing, and mathematical tools for induction and simplicity. 1983 edition. 352pp. 5 3/8 x 8 1/2. 0-486-47438-0

INTRODUCTION TO PROBABILITY THEORY WITH CONTEMPORARY APPLICATIONS, Lester L. Helms. Extensive discussions and clear examples, written in plain language, expose students to the rules and methods of probability. Exercises foster problem-solving skills, and all problems feature step-by-step solutions. 1997 edition. 368pp. 6 1/2 x 9 1/4. 0-486-47418-6

CHANCE, LUCK, AND STATISTICS, Horace C. Levinson. In simple, non-technical language, this volume explores the fundamentals governing chance and applies them to sports, government, and business. "Clear and lively ... remarkably accurate." – *Scientific Monthly.* 384pp. 5 3/8 x 8 1/2. 0-486-41997-5

FIFTY CHALLENGING PROBLEMS IN PROBABILITY WITH SOLUTIONS, Frederick Mosteller. Remarkable puzzlers, graded in difficulty, illustrate elementary and advanced aspects of probability. These problems were selected for originality, general interest, or because they demonstrate valuable techniques. Also includes detailed solutions. 88pp. 5 3/8 x 8 1/2. 0-486-65355-2

EXPERIMENTAL STATISTICS, Mary Gibbons Natrella. A handbook for those seeking engineering information and quantitative data for designing, developing, constructing, and testing equipment. Covers the planning of experiments, the analyzing of extreme-value data; and more. 1966 edition. Index. Includes 52 figures and 76 tables. 560pp. 8 3/8 x 11. 0-486-43937-2

STOCHASTIC MODELING: Analysis and Simulation, Barry L. Nelson. Coherent introduction to techniques also offers a guide to the mathematical, numerical, and simulation tools of systems analysis. Includes formulation of models, analysis, and interpretation of results. 1995 edition. 336pp. 6 1/8 x 9 1/4. 0-486-47770-3

INTRODUCTION TO BIOSTATISTICS: Second Edition, Robert R. Sokal and F. James Rohlf. Suitable for undergraduates with a minimal background in mathematics, this introduction ranges from descriptive statistics to fundamental distributions and the testing of hypotheses. Includes numerous worked-out problems and examples. 1987 edition. 384pp. 6 1/8 x 9 1/4. 0-486-46961-1

Browse over 9,000 books at www.doverpublications.com

Mathematics–Geometry and Topology

PROBLEMS AND SOLUTIONS IN EUCLIDEAN GEOMETRY, M. N. Aref and William Wernick. Based on classical principles, this book is intended for a second course in Euclidean geometry and can be used as a refresher. More than 200 problems include hints and solutions. 1968 edition. 272pp. 5 3/8 x 8 1/2.   0-486-47720-7

TOPOLOGY OF 3-MANIFOLDS AND RELATED TOPICS, Edited by M. K. Fort, Jr. With a New Introduction by Daniel Silver. Summaries and full reports from a 1961 conference discuss decompositions and subsets of 3-space; n-manifolds; knot theory; the Poincaré conjecture; and periodic maps and isotopies. Familiarity with algebraic topology required. 1962 edition. 272pp. 6 1/8 x 9 1/4.   0-486-47753-3

POINT SET TOPOLOGY, Steven A. Gaal. Suitable for a complete course in topology, this text also functions as a self-contained treatment for independent study. Additional enrichment materials make it equally valuable as a reference. 1964 edition. 336pp. 5 3/8 x 8 1/2.   0-486-47222-1

INVITATION TO GEOMETRY, Z. A. Melzak. Intended for students of many different backgrounds with only a modest knowledge of mathematics, this text features self-contained chapters that can be adapted to several types of geometry courses. 1983 edition. 240pp. 5 3/8 x 8 1/2.   0-486-46626-4

TOPOLOGY AND GEOMETRY FOR PHYSICISTS, Charles Nash and Siddhartha Sen. Written by physicists for physics students, this text assumes no detailed background in topology or geometry. Topics include differential forms, homotopy, homology, cohomology, fiber bundles, connection and covariant derivatives, and Morse theory. 1983 edition. 320pp. 5 3/8 x 8 1/2.   0-486-47852-1

BEYOND GEOMETRY: Classic Papers from Riemann to Einstein, Edited with an Introduction and Notes by Peter Pesic. This is the only English-language collection of these 8 accessible essays. They trace seminal ideas about the foundations of geometry that led to Einstein's general theory of relativity. 224pp. 6 1/8 x 9 1/4. 0-486-45350-2

GEOMETRY FROM EUCLID TO KNOTS, Saul Stahl. This text provides a historical perspective on plane geometry and covers non-neutral Euclidean geometry, circles and regular polygons, projective geometry, symmetries, inversions, informal topology, and more. Includes 1,000 practice problems. Solutions available. 2003 edition. 480pp. 6 1/8 x 9 1/4.   0-486-47459-3

TOPOLOGICAL VECTOR SPACES, DISTRIBUTIONS AND KERNELS, François Trèves. Extending beyond the boundaries of Hilbert and Banach space theory, this text focuses on key aspects of functional analysis, particularly in regard to solving partial differential equations. 1967 edition. 592pp. 5 3/8 x 8 1/2.
                    0-486-45352-9

INTRODUCTION TO PROJECTIVE GEOMETRY, C. R. Wylie, Jr. This introductory volume offers strong reinforcement for its teachings, with detailed examples and numerous theorems, proofs, and exercises, plus complete answers to all odd-numbered end-of-chapter problems. 1970 edition. 576pp. 6 1/8 x 9 1/4.   0-486-46895-X

FOUNDATIONS OF GEOMETRY, C. R. Wylie, Jr. Geared toward students preparing to teach high school mathematics, this text explores the principles of Euclidean and non-Euclidean geometry and covers both generalities and specifics of the axiomatic method. 1964 edition. 352pp. 6 x 9.   0-486-47214-0

Browse over 9,000 books at www.doverpublications.com

Mathematics–History

THE WORKS OF ARCHIMEDES, Archimedes. Translated by Sir Thomas Heath. Complete works of ancient geometer feature such topics as the famous problems of the ratio of the areas of a cylinder and an inscribed sphere; the properties of conoids, spheroids, and spirals; more. 326pp. 5 3/8 x 8 1/2. 0-486-42084-1

THE HISTORICAL ROOTS OF ELEMENTARY MATHEMATICS, Lucas N. H. Bunt, Phillip S. Jones, and Jack D. Bedient. Exciting, hands-on approach to understanding fundamental underpinnings of modern arithmetic, algebra, geometry and number systems examines their origins in early Egyptian, Babylonian, and Greek sources. 336pp. 5 3/8 x 8 1/2. 0-486-25563-8

THE THIRTEEN BOOKS OF EUCLID'S ELEMENTS, Euclid. Contains complete English text of all 13 books of the Elements plus critical apparatus analyzing each definition, postulate, and proposition in great detail. Covers textual and linguistic matters; mathematical analyses of Euclid's ideas; classical, medieval, Renaissance and modern commentators; refutations, supports, extrapolations, reinterpretations and historical notes. 995 figures. Total of 1,425pp. All books 5 3/8 x 8 1/2.
Vol. I: 443pp. 0-486-60088-2
Vol. II: 464pp. 0-486-60089-0
Vol. III: 546pp. 0-486-60090-4

A HISTORY OF GREEK MATHEMATICS, Sir Thomas Heath. This authoritative two-volume set that covers the essentials of mathematics and features every landmark innovation and every important figure, including Euclid, Apollonius, and others. 5 3/8 x 8 1/2.
Vol. I: 461pp. 0-486-24073-8
Vol. II: 597pp. 0-486-24074-6

A MANUAL OF GREEK MATHEMATICS, Sir Thomas L. Heath. This concise but thorough history encompasses the enduring contributions of the ancient Greek mathematicians whose works form the basis of most modern mathematics. Discusses Pythagorean arithmetic, Plato, Euclid, more. 1931 edition. 576pp. 5 3/8 x 8 1/2.
0-486-43231-9

CHINESE MATHEMATICS IN THE THIRTEENTH CENTURY, Ulrich Libbrecht. An exploration of the 13th-century mathematician Ch'in, this fascinating book combines what is known of the mathematician's life with a history of his only extant work, the Shu-shu chiu-chang. 1973 edition. 592pp. 5 3/8 x 8 1/2.
0-486-44619-0

PHILOSOPHY OF MATHEMATICS AND DEDUCTIVE STRUCTURE IN EUCLID'S ELEMENTS, Ian Mueller. This text provides an understanding of the classical Greek conception of mathematics as expressed in Euclid's Elements. It focuses on philosophical, foundational, and logical questions and features helpful appendixes. 400pp. 6 1/2 x 9 1/4. 0-486-45300-6

BEYOND GEOMETRY: Classic Papers from Riemann to Einstein, Edited with an Introduction and Notes by Peter Pesic. This is the only English-language collection of these 8 accessible essays. They trace seminal ideas about the foundations of geometry that led to Einstein's general theory of relativity. 224pp. 6 1/8 x 9 1/4. 0-486-45350-2

HISTORY OF MATHEMATICS, David E. Smith. Two-volume history – from Egyptian papyri and medieval maps to modern graphs and diagrams. Non-technical chronological survey with thousands of biographical notes, critical evaluations, and contemporary opinions on over 1,100 mathematicians. 5 3/8 x 8 1/2.
Vol. I: 618pp. 0-486-20429-4
Vol. II: 736pp. 0-486-20430-8